Iron Intake and Human Health

Iron Intake and Human Health

Editor

Gladys Oluyemisi Latunde-Dada

MDPI

Basel • Beijing • Wuhan • Barcelona • Belgrade • Novi Sad • Cluj • Manchester

Editor
Gladys Oluyemisi
Latunde-Dada
King's College London
London, UK

Editorial Office
MDPI
St. Alban-Anlage 66
4052 Basel, Switzerland

This is a reprint of articles from the Special Issue published online in the open access journal *Nutrients* (ISSN 2072-6643) (available at: https://www.mdpi.com/journal/nutrients/special_issues/Iron_Human_Health).

For citation purposes, cite each article independently as indicated on the article page online and as indicated below:

Lastname, A.A.; Lastname, B.B. Article Title. *Journal Name* **Year**, *Volume Number*, Page Range.

ISBN 978-3-0365-9640-2 (Hbk)
ISBN 978-3-0365-9641-9 (PDF)
doi.org/10.3390/books978-3-0365-9641-9

Contents

About the Editor

Gladys Oluyemisi Latunde-Dada

Dr Gladys Oluyemisi Latunde-Dada is a Senior Lecturer in the Department of Nutritional Sciences, King's College London, UK. Her research interests are the molecular biology of iron metabolism and iron nutrition. She was involved in the characterization of iron metabolism genes and proteins in the mammalian system by the Iron Research Group at King's. The group has contributed immensely to the global understanding of iron transport and metabolism. Her interests in iron metabolism research include studies on the Anemia of Chronic Kidney Disease and the evaluation of iron and cancer interactions in cell culture models. In addition, she has also researched the bioavailability of iron nano-compounds and amino acid iron chelates in cell culture and animal studies. She is a member of a collaborative group that studies the bioavailability of iron and zinc from wheat in human subjects. She has recently commenced studies on ferroptosis, a form of programmed cell death akin to apoptosis, which is induced by iron and lipid peroxidation. She is a Registered Nutritionist and a member of the Nutrition Society, the American Society for Nutrition, the Biochemical Society, the European Iron Club, and the International BIOIRON Society.

nutrients MDPI

Editorial

Iron Intake and Human Health

Gladys O. Latunde-Dada

Department of Nutritional Sciences, School of Life Course and Population Sciences, King's College London, London SE1 9NH, UK; yemisi.latunde-dada@kcl.ac.uk

Iron deficiency anemia (IDA) is a global nutritional disorder affecting large population groups in varying magnitudes in different countries [1]. Such groups include vulnerable growing children, adolescent girls, pregnant and lactating women, and the aged. The physiological challenges of IDA, particularly those impacting growth and reproductive purposes, pose demands for increased daily iron requirements for these demographics. The daily iron needs of these vulnerable groups are commonly unmet because of inadequate iron intake, poor bioavailability, low absorption of iron due to the presence of inhibitors in the diet, or chronic blood loss, amongst others [2]. Consequently, evidence abounds regarding the deleterious and debilitating consequences of IDA on infants' cognition, brain function, mental capability, work performance, and pregnancy outcomes [3,4]. Strategies to combat IDA include fortification of foods, iron supplementation to targeted groups, and attempts to explore and exploit food-based strategies to enhance iron absorption. For example, recent publications have revealed iron bioavailability data on novel food products, new oral iron supplements [5,6], and iron-biofortified foods that are safe and less toxic to the gut. Moreover, food processing procedures such as micro-milling and food encapsulation are employed to influence luminal bioaccessibility and iron absorption from foods [7]. Furthermore, gut microflora have emerged as important entities that could modify food matrices while secreting metabolites that might modulate iron absorption in the proximal gut region [8].

Besides the common causes of IDA, inflammation places a significant burden on systemic iron metabolism and the iron status of individuals with varying conditions. IDA in developing countries is often associated with inflammation and infections. Indeed, inflammation is a key regulator of hepcidin, the peptide hormone that modulates iron absorption and iron homeostasis [9]. Hepcidin expression is elevated in response to proinflammatory cytokines via a STAT3 transcriptional signaling pathway. This causes reduced iron absorption, since ferroportin, the iron efflux protein, is ubiquitinated by hepcidin [10]. Hence, acute or low-grade inflammation therefore contributes to the incidence of anemia due to the inhibition of iron absorption in the gut. In this issue, Htet [11] reported, in a randomized, double-blinded, placebo-controlled trial, the effect of sub-clinical inflammation (SCI) on the absorption of iron supplements in anemic adolescent schoolgirls in the Ayeyarwady region of Myanmar. The study also investigated the effect of combining vitamin A or folic acid with iron supplements given to the subjects with SCI. The study found that the efficiency of iron supplements, as expected, was inhibited by SCI and vitamin A, which mitigated their effect. Hence, the anti-inflammatory function of vitamin A neutralizes SCI to facilitate enhanced iron absorption. Moreover, previous studies have indicated the beneficial effect of vitamin A on the mobilization of stored iron for the enhancement of erythropoietin synthesis [12,13].

The prevalence of IDA is also associated with genetic variability in human populations [14]. For example, human genome-wide association studies (GWAS) have identified the association of multiple variants of the transmembrane protease serine 6 (TMPRSS6) enzyme with abnormal hematological biomarkers. As a negative modulator of hepcidin expression, TMPRSS6 down-regulates the transcription of hepcidin via the cleavage of

Citation: Latunde-Dada, G.O. Iron Intake and Human Health. *Nutrients* **2024**, *16*, 206. https://doi.org/10.3390/nu16020206

Received: 2 December 2023
Accepted: 28 December 2023
Published: 8 January 2024

hemojuvelin. Mutations in the TMPRSS6 gene lead to loss of function of the matriptase-2 protein, which induces increased hepcidin levels in the liver. Hence, elevated hepcidin levels, by degradation of ferroportin, lead to severe microcytic anemia due to reduced iron absorption in the gastrointestinal tract [15]. Shinta et al. [16] compared the association of two common TMPRSS6 SNPs (rs855791 and rs4820268) with iron intake and its effects on the prevalence of IDA in children under two years of age. While the two TMPRSS6 SNP variants were associated with lower serum ferritin concentrations, the association of IDA with iron intake was significantly greater in these children [16]. In contrast, however, De Oliveira Mota et al. [17] reported in the current issue that a red meat intake of 100 g/d would theoretically not reduce the estimated burden of IDA, particularly in menstruating females in population groups from France. This finding was obtained from a predictive mathematical model that quantitatively determined the health burden of IDA in terms of disability-adjusted life years (DALY) based on the age and gender of the subjects as well as their iron requirements, intake, and status in France. In another study in this issue [18], non-heme iron and total iron intake, rather than heme iron intake, were positively associated with MetS (waist circumference, reduced HDL-C, and elevated triglycerides and blood pressure) in the adult male population in China. An association between iron intake and MetS risk was reported only in men, but not in women, possibly because of reduced intake and increased loss of iron from the body due to the menstrual cycle in the latter. It should be noted, however, that the dietary iron intake was calculated from a 3-day diet record with inherent confounding issues. Nevertheless, evidence abounds in the literature [19–21] on the complex relationship between iron metabolism and the risk of MetS. The mechanism underlying these interactions is a subject of increasing research and may be linked to ferroptosis.

Moreover, the dysregulation of iron metabolism has been associated with the derangement of body tissue composition, energy expenditure, and metabolic syndrome. Moreno-Fernandez et al. [22] reported that Fe deficiency resulted in significant reductions in lean tissue mass and body fat, as well as inducing low energy expenditure in mice. While the levels of the triiodothyronine, thyroxine, and ghrelin hormones decreased in the anemic mice, glucose-dependent insulinotropic polypeptide (GIP), glucagon, insulin, corticosterone, and adrenocorticotropic hormone levels were elevated. Low oxygen consumption and hypoxia in the anemic mice may possibly have impacted the iron-dependent enzymes that are involved in ATP synthesis and the resultant weight loss in the mice, as was also reported by Schneider et al. [23].

The health burden of IDA also encompasses debilitating consequences on an infant's growth, fatigue, and mental function. IDA in fetuses and babies during the pre- and post-natal periods could cause impaired memory, reduced attentiveness, and cognitive malfunction, amongst other issues [24]. Moreno-Fernandez et al. [25] performed a systematic literature review on the effect of iron status on the growth and developmental indices of premature infants. Iron demands are high during early growth, and it has been proposed to give iron supplements to vulnerable infants, particularly those with low birth weights. The study [24] also showed that low iron overload status in pre-term infants was more associated with lower birth weights than those with normal iron status. Hence, there is a delicate balance between the optimal iron supplementation dose to avert the negative consequences of IDA and the levels required to prevent organ damage due to iron overload. For example, a study [26] reported that some pregnant Norwegian women had iron supplementation beyond the daily recommended dose, which could cause adverse toxic effects to both the mothers and newborn babies. Nonetheless, a large-scale prospective study in the current issue revealed no deleterious consequences of iron supplementation administered during pregnancy and the postpartum period to the risk of developing Type 1 diabetes (T1D) in babies [27]. This Danish study concluded that maternal supplemental iron intake was possibly protective against T1D, rather than being a risk factor. However, there are instances when unabsorbed supplemented iron can induce the formation of toxic reactive oxygen species that cause inflammation or enhance the proliferation of iron-dependent pathogenic microbes in the distal gut region [28]. The deleterious consequence of high concentrations of supplemental iron in the colon has indeed been associated with gastrointestinal inflammation and microbial dysbiosis [29]. Iron chelation to starve and sequester iron from pathogenic microbes has been advocated as a panacea to maintain good gut health. Based on this premise, Horniblow et al. [30] investigated the safety and tolerability of a biopolymeric alginate as an iron chelator on hematological biomarkers and microbiome composition in healthy subjects. While the biopolymeric compound was safe and well tolerated, there were no changes to microbiota populations, and the iron-chelating ability of alginate was presumed to be confounded, possibly by the presence of other dietary chelators in the luminal food matrix content. Further published research by the same authors [31] showed that another biopolymer compound, namely, lignin, chelated iron and restricted its availability to detrimental proteobacteria while promoting the growth of favored beneficial bacteroides. Hence, lignin could be used to control microbial dysbiosis and inflammation associated with gut disorders. Regarding the public health significance of IDA, strategies to alleviate its effects include food fortification programs that are adopted by different countries and iron supplementation to specific vulnerable groups [32,33] across different age groups. Iron compounds used for supplementation include ferrous sulfate, elemental iron, ferric pyrophosphate, ferrous sulfate, ferric pyrophosphate, ferric ammonium citrate, ferrous sulfate, ferric pyrophosphate, fumarate, gluconate, and ferric ammonium citrate, among several others. Nevertheless, some of these iron salts cause gastrointestinal side effects such as nausea, constipation, diarrhea, and inflammation, as discussed earlier. Therefore, the initiative to synthesize novel iron salts that are safe, tolerable, and of high bioavailability is an ongoing challenge. Recently, newly synthesized nanoparticulate, ligand-modified Fe(III) polyoxo-hydroxide [34], nanoparticulate Fe pyrophosphates [35], sucrosomal iron [5], and iron multi-amino acid chelate (IMAAC) [6] have been added to the list of iron supplements to be used to prevent and treat IDA. In line with this development, Naviglio et al. [36] reported a method by which an iron (II) citrate complex was synthesized using iron filings and citric acid. The compound, $FeC_6H_6O_7{\cdot}H_2O$, which contains Fe(II), was purified with an array of analytical techniques, and it has been proposed to be available as an iron supplement for commercial purposes. For these reasons, it is imperative to evaluate and compare the in vitro bioaccessibility and in vivo bioavailability of $FeC_6H_6O_7{\cdot}H2O$ with the gold-standard $FeSO_4$. Fe(II) is the form of iron that is absorbed by divalent metal transporter 1, DMT1, the metal transporter in the duodenum of the gastrointestinal tract. Yu et al. [37] investigated the transport of iron from ferrous bis-glycinate (Fe-Gly) in CRISPR-Cas 9 DMT1-knockout Caco-2 cells. These authors reported that iron from Fe-Gly was mainly absorbed by Caco-2 cells via DMT1, and that iron-regulated transporter (IRT)-like protein 14 (Zip14) expression was induced to compensate for iron absorption when DMT1 was limiting.

In summary, the papers in this issue, which were published originally in *Nutrients*, explored the significance of IDA, spanning across its causes and consequences. Furthermore, some approaches and interventions to prevent its incidence in human populations were highlighted and discussed. The health burden of IDA includes debilitating effects on energy metabolism, inflammation, and infection. Strategies to synthesize iron supplements of high bioavailability, with high redox inert and without adverse effects in the gastrointestinal tract, will be of immense benefit to the drive to prevent, manage, and treat IDA globally.

Conflicts of Interest: The authors declare no conflict of interest.

References

1. World Health Organization. Micronutrient Deficiencies—Iron Deficiency Anaemia. 2014. Available online: http://www.who.int/nutrition/topics/ida/en/# (accessed on 2 November 2014).
2. Aspuru, K.; Villa, C.; Bermejo, F.; Herrero, P.; García López, S. Optimal management of iron deficiency anemia due to poor dietary intake. *Int. J. Gen. Med.* **2011**, *4*, 741–750. [PubMed]
3. Munoz, P.; Humeres, A. Iron deficiency on neuronal function. *Biometals* **2012**, *25*, 825–835. [CrossRef] [PubMed]

4. Rufer, E.S.; Tran, T.D.; Attridge, M.M.; Andrzejewski, M.E.; Flentke, G.R.; Smith, S.M. Adequacy of maternal iron status protects against behavioral, neuroanatomical, and growth deficits in fetal alcohol spectrum disorders. *PLoS ONE* **2012**, *7*, e47499. [CrossRef] [PubMed]
5. Elli, L.; Ferretti, F.; Branchi, F.; Tomba, C.; Lombardo, V.; Scricciolo, A.; Doneda, L.; Roncoroni, L. Sucrosomial Iron Supplementation in Anemic Patients with Celiac Disease Not Tolerating Oral Ferrous Sulfate: A Prospective Study. *Nutrients* **2018**, *10*, 330. [CrossRef] [PubMed]
6. Kajarabille, N.; Brown, C.; Cucliciu, A.; Thapaliya, G.; Latunde-Dada, G.O. Bioavailability of iron multi-amino acid chelate preparation in mice and human duodenal HuTu 80 cells. *Br. J. Nutr.* **2017**, *117*, 767–774. [CrossRef] [PubMed]
7. Latunde-Dada, G.O.; Li, X.; Parodi, A.; Edwards, C.H.; Ellis, P.R.; Sharp, P.A. Micromilling enhances iron bioaccessibility from wholegrain wheat. *J. Agric. Food Chem.* **2014**, *62*, 11222–11227. [CrossRef]
8. Kortman, G.A.M.; Reijnders, D.; Swinkels, D.W. Oral iron supplementation: Potential implications for the gut microbiome and metabolome in patients with CKD. *Hemodial Int.* **2017**, *21* (Suppl. 1), S28–S36. [CrossRef] [PubMed]
9. Silvestri, L.; Nai, A.; Dulja, A.; Pagani, A. Hepcidin and the BMP-SMAD pathway: An unexpected liaison. *Vitam. Horm.* **2019**, *110*, 71–99.
10. Nemeth, E.; Ganz, T. Hepcidin-Ferroportin Interaction Controls Systemic Iron Homeostasis. *Int. J. Mol. Sci.* **2021**, *22*, 6493. [CrossRef]
11. Htet, M.K.; Fahmida, U.; Dillon, D.; Akib, A.; Utomo, B.; Thurnham, D.I. Is Iron Supplementation Influenced by Sub-Clinical Inflammation?: A Randomized Controlled Trial Among Adolescent Schoolgirls in Myanmar. *Nutrients* **2019**, *11*, 918. [CrossRef]
12. Zimmermann, M.B.; Biebinger, R.; Rohner, F.; Dib, A.; Zeder, C.; Hurrell, R.F.; Chaouki, N. Vitamin A supplementation in children with poor vitamin A and iron status increases erythropoietin and hemoglobin concentrations without changing total body iron. *Am. J. Clin. Nutr.* **2006**, *84*, 580–586. [CrossRef] [PubMed]
13. Semba, R.; Bloem, M. The anemia of vitamin A deficiency: Epidemiology and pathogenesis. *Eur. J. Clin. Nutr.* **2002**, *56*, 271–281. [CrossRef] [PubMed]
14. Tanaka, T.; Roy, C.N.; Yao, W.; Matteini, A.; Semba, R.D.; Arking, D.; Walston, J.D.; Fried, L.P.; Singleton, A.; Guralnik, J.; et al. A genome-wide association analysis of serum iron concentrations. *Blood J.* **2010**, *115*, 94–96. [CrossRef] [PubMed]
15. Gichohi-Wainaina, W.N.; Towers, G.W.; Swinkels, D.W.; Zimmermann, M.B.; Feskens, E.J.; Melse-Boonstra, A. Inter-ethnic differences in genetic variants within the transmembrane protease, serine 6 (TMPRSS6) gene associated with iron status indicators: A systematic review with meta-analyses. *Genes Nutr.* **2014**, *10*, 442. [CrossRef] [PubMed]
16. Shinta, D.; Asmarinah; Adhiyanto, C.; Htet, M.K.; Fahmida, U. The Association of TMPRSS6 Gene Polymorphism and Iron Intake with Iron Status among Under-Two-Year-Old Children in Lombok, Indonesia. *Nutrients* **2019**, *11*, 878. [CrossRef] [PubMed]
17. Mota, J.O.; Tounian, P.; Guillou, S.; Pierre, F.; Membré, J.M. Estimation of the Burden of Iron Deficiency Anemia in France from Iron Intake: Methodological Approach. *Nutrients* **2019**, *11*, 2045. [CrossRef] [PubMed]
18. Zhu, Z.; Wu, F.; Lu, Y.; Wu, C.; Wang, Z.; Zang, J.; Guo, C.; Jia, X.; Yao, J.; Peng, H.; et al. Total and Nonheme Dietary Iron Intake Is Associated with Metabolic Syndrome and Its Components in Chinese Men and Women. *Nutrients* **2018**, *10*, 1663. [CrossRef]
19. Li, Y.; Zhao, L.; Yu, D.; Wang, Z.; Ding, G. Metabolic syndrome prevalence and its risk factors among adults in China: A nationally representative cross-sectional study. *PLoS ONE* **2018**, *13*, e199293. [CrossRef]
20. Kilani, N.; Vollenweider, P.; Waeber, G.; Marques-Vidal, P. Iron metabolism and incidence of metabolic syndrome. *Nutr. Metab. Cardiovasc. Dis.* **2015**, *25*, 1025–1032. [CrossRef]
21. Dos Santos Vieira, D.; Hermes Sales, C.; Galvão Cesar, C.; Marchioni, D.; Fisberg, R. Influence of Haem, Non-Haem, and Total Iron Intake on Metabolic Syndrome and Its Components: A Population-Based Study. *Nutrients* **2018**, *10*, 314. [CrossRef]
22. Moreno-Fernandez, J.; Díaz-Castro, J.; Alférez, M.J.M.; López-Aliaga, I. Iron Deficiency and Neuroendocrine Regulators of Basal Metabolism, Body Composition and Energy Expenditure in Rats. *Nutrients* **2019**, *11*, 631. [CrossRef] [PubMed]
23. Schneider, J.M.; Fujii, M.L.; Lamp, C.L.; Lönnerdal, B.; Dewey, K.G. The use of multiple logistic regression to identify risk factors associated with anemia and iron deficiency in a convenience sample of 12–36-mo-old children from low-income families. *Am. J. Clin. Nutr.* **2008**, *87*, 614–620. [CrossRef] [PubMed]
24. Moreno-Fernandez, J.; Ochoa, J.J.; Latunde-Dada, G.O.; Diaz-Castro, J. Iron Deficiency and Iron Homeostasis in Low Birth Weight Preterm Infants: A Systematic Review. *Nutrients* **2019**, *11*, 1090. [CrossRef] [PubMed]
25. Reid, B.M.; Georgieff, M.K. The Interaction between Psychological Stress and Iron Status on Early-Life. *Nutrients* **2023**, *15*, 3798. [CrossRef] [PubMed]
26. Thorsen, S.U.; Halldorsson, T.I.; Bjerregaard, A.A.; Olsen, S.F.; Svensson, J. Maternal and Early Life Iron Intake and Risk of Childhood Type 1 Diabetes: A Danish Case-Cohort Study. *Nutrients* **2019**, *11*, 734. [CrossRef] [PubMed]
27. Størdal, K.; McArdle, H.J.; Hayes, H.; Tapia, G.; Viken, M.K.; Lund-Blix, N.A.; Haugen, M.; Joner, G.; Skrivarhaug, T.; Mårild, K.; et al. Prenatal iron exposure and childhood type 1 diabetes. *Sci. Rep.* **2018**, *8*, 9067. [CrossRef] [PubMed]
28. Mahalhal, A.; Williams, J.M.; Johnson, S.; Ellaby, N.; Duckworth, C.A.; Burkitt, M.D.; Liu, X.; Hold, G.L.; Campbell, B.J.; Pritchard, D.M. Oral iron exacerbates colitis and influences the intestinal microbiome. *PLoS ONE* **2018**, *13*, e0202460. [CrossRef]
29. Tang, M.; Frank, D.N.; Hendricks, A.E.; Ir, D.; Esamai, F.; Liechty, E.; Hambidge, K.M.; Krebs, N.F. Iron in micronutrient powder promotes an unfavorable gut microbiota in Kenyan infants. *Nutrients* **2017**, *9*, 776. [CrossRef]
30. Horniblow, R.D.; Mistry, P.; Quraishi, M.N.; Beggs, A.D.; Van de Wiele, T.; Iqbal, T.H.; Tselepis, C. The Safety and Tolerability of a Potential Alginate-Based Iron Chelator; Results of A Healthy Participant Study. *Nutrients* **2019**, *11*, 674. [CrossRef]

31. Horniblow, R.D.; Pathak, P.; Eshrati, M.; Latunde-Dada, G.O.; Tselepis, C. Intestinal iron bio-accessibility changes by Lignin and the subsequent impact on cell metabolism and intestinal microbiome communities. *Food Funct.* **2023**, *14*, 3673–3685. [CrossRef]
32. Gera, T.; Sachdev, H.S.; Boy, E. Effect of iron-fortified foods on hematologic and biological outcomes: Systematic review of randomized controlled trials. *Am. J. Clin. Nutr.* **2012**, *96*, 309–324. [CrossRef]
33. Tiwari, A.K.; Mahdi, A.A.; Chandyan, S.; Zahra, F.; Godbole, M.M.; Jaiswar, S.P.; Srivastava, V.K.; Negi, M.P.S. Oral iron supplementation leads to oxidative imbalance in anemic women: A prospective study. *Clin. Nutr.* **2011**, *30*, 188–193. [CrossRef]
34. Hilty, F.M.; Arnold, M.; Hilbe, M.; Teleki, A.; Knijnenburg, J.T.N.; Ehrensperger, F.; Hurrell, R.F.; Pratsinis, S.E.; Langhans, W.; Zimmermann, M.B. Iron from nanocompounds containing iron and zinc is highly bioavailable in rats without tissue accumulation. *Nat. Nanotechnol.* **2010**, *5*, 374–380. [CrossRef]
35. Pereira, D.I.A.; Bruggraber, S.F.A.; Faria, N.; Poots, L.K.; Tagmount, M.A.; Aslam, M.F.; Frazer, D.M.; Vulpe, C.D.; Anderson, G.J.; Powell, J.J. Nanoparticulate iron(III) oxo-hydroxide delivers safe iron that is well absorbed and utilised in humans. *Nanomedicine* **2014**, *10*, 1877–1886. [CrossRef]
36. Naviglio, D.; Salvatore, M.M.; Limatola, M.; Langella, C.; Faralli, S.; Ciaravolo, M.; Andolfi, A.; Salvatore, F.; Gallo, M. Iron (II) Citrate Complex as a Food Supplement: Synthesis, Characterization and Complex Stability. *Nutrients* **2018**, *10*, 1647. [CrossRef]
37. Yu, X.; Chen, L.; Ding, H.; Zhao, Y.; Feng, J. Iron Transport from Ferrous Bisglycinate and Ferrous Sulfate in DMT1-Knockout Human Intestinal Caco-2 Cells. *Nutrients* **2019**, *11*, 485. [CrossRef] [PubMed]

nutrients

MDPI

Article

Is Iron Supplementation Influenced by Sub-Clinical Inflammation?: A Randomized Controlled Trial Among Adolescent Schoolgirls in Myanmar

Min Kyaw Htet [1,2,3,*], Umi Fahmida [1], Drupadi Dillon [1,4], Arwin Akib [4], Budi Utomo [5] and David I Thurnham [6]

1 South East Asian Ministers of Education Organization, Regional Center for Food and Nutrition (SEAMEO RECFON) Pusat Kajian Gizi Regional, Universitas Indonesia, Jakarta 10430, Indonesia; umifahmida@gmail.com or ufahmida@seameo-recfon.org (U.F.); ddrupadi@hotmail.com (D.D.)
2 Township Health Department, Ministry of Health Myanmar, Hakha 03011, Chin State, Myanmar
3 Sydney School of Public Health, Sydney Medical School, The University of Sydney, Sydney NSW 2006, Australia
4 Faculty of Medicine, University of Indonesia, Jakarta 10430, Indonesia; aralpur@yahoo.com
5 Faculty of Public Health, University of Indonesia, Depok 16424, Indonesia; budi.utomo.ui@gmail.com
6 Northern Ireland Centre for Food and Health, School of Biomedical Science, University of Ulster, Coleraine BT52 1SA, UK; di.thurnham@ulster.ac.uk
* Correspondence: kyawhtet@gmail.com; Tel.: +62-857-2726-2672 or +95-925-023-7575

Received: 14 March 2019; Accepted: 16 April 2019; Published: 24 April 2019

Abstract: Iron absorption was impaired in the presence of sub-clinical inflammation (SCI) and might hamper the effect of iron supplementation. The purpose of the study was to identify the influence of SCI on iron supplementation. A randomized, double-blinded, placebo-controlled experimental study was conducted among anaemic adolescent schoolgirls in Ayeyarwady region, Myanmar. A total of 402 schoolgirls were recruited from six schools screened from 1269 girls who were assigned into one of four groups: Folate group (2.5 mg of folate), Vitamin A group (15,000 IU of vitamin), Iron folate group (60 mg elemental iron and folate) and Iron, and vitamin A and folate group. Supplementation was done once a week for 12 weeks. Iron, vitamin A and inflammation were measured at the baseline, middle and endline. Changes in serum ferritin and body iron were significantly higher in the IFA and IFA + vitA among those without SCI. There was interaction between vitamin A and SCI on Hb changes. Analysis of GLM repeated measure showed interactions between treatment and SCI for hemoglobin and serum transferrin receptor. Those treated with vitamin A had better outcomes when there was SCI. Inflammation accompanied a negative effect on iron supplementation and vitamin A improved efficacy of iron supplementation in the presence of SCI.

Keywords: anemia; iron deficiency; Myanmar; sub-clinical inflammation

1. Introduction

Despite the progress in anemia control over the last decades, iron deficiency and anemia still remain major nutritional problem in many parts of the world [1]. Fortification and food based approaches are promising strategies to overcome the problem but iron supplementation remains important since it is the most effective strategy for the areas where iron deficiency is common and anemia is a major public health problem [2,3]. Anemia and iron deficiency are prevalent in developing countries and are often accompanied by infections and sub-clinical infection (SCI). The interaction between infections and nutritional status of individuals is well recognized and in recent years the relationship between the host iron status and infections has become a focus of research interest [4]. Not all studies have shown iron supplementation to be effective in reducing anemia and it may even cause deleterious health outcomes in the area where infections, particularly malaria, are common as, for example, in the Pemba trial [5]. Subsequently to that study, WHO in 2006, recommended targeted and supervised supplementation in malaria endemic areas only to iron deficient or severely anemic young children [6]. However, in the low resource settings where anemia and iron deficiency are prevalent, it is not always feasible to assess the iron status of the population before giving supplementation.

Understanding of the interaction between infections and iron status is gradually improving yet the work remains challenging. The function of the newly-discovered, iron-regulatory hormone hepcidin explains why anemia commonly occurs in the presence of chronic diseases and infections [7]. While the debate is ongoing for iron supplementation in malaria-endemic areas, little attention is paid to the importance of sub-clinical inflammation (SCI) in relation to the effectiveness of iron supplementation. SCI can exist in the apparently-healthy population and unless taken into account, its influence on anemia will be overlooked [8]. Studies showed that SCI can interfere with the assessment of micronutrient status especially indicators of iron and vitamin A status [9–11]. However, there have been no attempts to investigate the influence of SCI on the effectiveness of iron supplementation. Studies have shown additional benefits in reducing anemia by combining iron and vitamin A, but none have considered the role of SCI or anti-inflammatory properties of vitamin A on the mechanisms involved.

We hypothesized that the effect of iron supplementation on anemia may not be optimal in the presence of SCI and explored the interactive effect of combined iron and vitamin A supplementation and SCI. The study was registered at the clinical trial registration system with the identifier ClinicalTrials.gov ID: NCT 01198574.

2. Materials and Methods

Study setting and population: Background information on the subjects and study area are described elsewhere [12]. The study was conducted in Nyaung Done Townhsip, a peri-urban area in Ayeyarwady division, the delta region of Myanmar located about 3 h drive, (~40 km) distance from the former capital city Yangon.

The subjects were 8 to 10 grade post-menarcheal, adolescent school girls and were recruited in July 2010. Screening for anemia, malaria and haemoglobinopathies was done among the schoolgirls (n = 1269) from six schools in the study area during the recruitment (5). Postmenarcheal schoolgirls with anemia (Hb < 12.0 g/dL), not suffering from any major illness or known disease at the time of recruitment, were invited to join the study. Those with severe anemia (Hb < 7.0 g/dL) or those who took multivitamin supplements regularly for past three months were not eligible (2). The girls with severe anemia were treated with iron tablets and not included in the study. Deworming with a single dose of albendazole (400 mg) was done in these girls one week before the blood collection.

Screening for anemia, hemolytic anemia and malaria infection was done in 6 schools among 8 to 10 grade schoolgirls and details were reported elsewhere [13]. The sample size was calculated in order to detect a mean hemoglobin difference of 5 g/L between two groups, with 95% power at 5% level of significance based on a previous study [14]. The sample size was 73 for each group and allowed for a dropout rate of 20%. To take account of inflammation, the sample was increased to 100 and a total of 402 subjects for four interventions were recruited. Deworming with a single dose of Albendazole (400 mg) was done to these girls one week before the blood collection.

Study design and Intervention: The study was a randomized, double blinded, placebo controlled trial and the subjects in all 6 schools were randomly assigned into one of the four treatment groups using random allocation software with the block size of 8. The block randomization was done by computer generated randomization and was done by a statistician. The randomization list was kept by someone who was not involved in the study. Allocation concealment was done by enclosing the assignment of treatment in sequentially numbered, opaque, sealed envelopes. This allocation concealment was done to prevent selection bias and it was done at the time of enrolment to protect the assignment sequences. The study personnel and participants were blinded to treatment assignment for the duration of the study.

There were four treatment groups involved in the study and each group received two tablets as follows: IFA group (60 mg iron with 2.5 mg folate and vitamin A placebo), VA group (VA 15,000 IU and placebo iron i.e., 2.5 mg folate), IFA+VA group (60 mg iron with 2.5 mg folate and VA 15,000 IU), and placebo group (FA) (2.5 mg folate and vitamin A placebo). The intervention groups were equivalent to 2 by 2 design with iron and vitamin A, but with all the groups receiving folate to ensure that anemia due to folate deficiency was controlled for. Supplementation was given once a week for 12 weeks and the weekly dosage of iron folate corresponded to previous studies [15] and that of vitamin A was calculated based on RDA (Recommended Daily Allowance). Supplements were prepared for the study by Kimia Farma Pharmaceutical Factory, Indonesia and placebos were made identical in appearance to the supplements and were coded at the factory. All the personnel and investigators were blinded and de-blinding of the code was done only after the preliminary data analyses were done.

Compliance: Supplementation was done once a week for twelve weeks under close supervision of school teachers. During the intervention, principal investigator and field assistants visited the schools every week to monitor any side effects, to check the compliance, and to collect empty supplement containers and provide the supplements for the upcoming week.

Data Collection: At baseline, participants were administered a standard interview using structured questionnaire. Anthropometric assessments were carried out by two trained persons. Body weight was measured by flat electronic weighing scale (SECA 874, Hamburg, Germany) to the nearest 0.1 kg and height measured by measuring tape (Microtoise SECA 206, Hamburg, Germany) to the nearest 0.1 cm. Anthropometric nutritional status was assessed using z scores and WHO anthroplus software was used for the analysis [16]. Venous blood samples were collected three times at baseline, midline, and endline (week 0, week 6 and after week 12) of the study. In the presence of a physician, phlebotomy was done by experienced nurses and 3 mL of non fasting venous blood was taken from each subject in the morning school session 8:00 to 12:00 hr. The blood samples were collected into a non-heparinised vacuette as well as into an EDTA tube and transported in ice-cooled containers to the laboratory of Department of Medical Research, Lower Myanmar (DMR-LM) within 4 h.

2.1. Ethics and Registration

Ethical approval of the study was granted from Faculty of Medicine, University of Indonesia (128/PT 02.FK/ETIK/2010) and also from Department of Medical Research, Lower Myanmar (18/Ethics 2010 DMR-Lower Myanmar). The study was registered at the clinical trial registration system at ClinicalTrials.gov (identifier ID: NCT 01198574).

Laboratory analysis: Hemoglobin was measured by cyanmethemoglobin method at Nutrition Research Division, DMR-LM. The quality control for the Hb assessment was done with control samples of known values from National Health Laboratory on daily basis and the coefficient of variation of the assessment was below 5.0%. Blood samples were centrifuged at 1500× g for 10 min and serum samples were kept in polyethylene vials at –20C before further analyses were done. These samples were kept frozen in dry ice during the shipment to the laboratory of South East Asian Ministers of Education Organization-Regional Center for Food and Nutrition (SEAMEO-RECFON), University of Indonesia, Jakarta, Indonesia for further analyses. Details of these analyses were described in a previous paper [12]. Briefly, an in-house sandwich enzyme-linked immunosorbent assay was used to measure serum ferritin (SF), transferrin receptor (sTfR), retinol binding protein (RBP), C-reactive protein (CRP), and α1-acid glycoprotein (AGP) on 200 µL serum [17]. Body iron store was calculated using the method of Cook et al. [18]. Serum retinol concentrations were measured by high performance liquid chromatography [19]. Serum folate concentrations were measured using microbiological assays according to the methods of O'Broin and Kelleher [20].

Anemia was defined as haemoglobin <12.0 g/dL [21], iron deficiency (ID) when SF < 15 µg/L and /or sTfR was >8.5 mg/L and IDA by the occurrence of ID and anemia [11,22]. Low vitamin A status was defined when serum retinol concentrations were <1.05 µmol/L) [23]. SCI was defined when serum CRP>5mg/L and/or AGP concentrations were >1g/L at any time throughout the study (baseline, midline, endline) [9,10].

2.2. Statistical Analysis

Statistical analyses were performed using statistical software package SPSS (SPSS; Chicago, IL USA) Version 15.0 for windows. Normality of distribution of the variables was checked with Komogorov-Smirnov test and serum ferritin, sTfR, and serum retinol were log transformed to better approximate a normal distribution. Data were presented as means ± SD or geometric means ± SD.

The baseline biochemical status of subjects among 4 groups was compared by one-way analysis of variance (ANOVA). Posthoc analysis using LSD was applied when ANOVA was statistically significant. The prevalence of iron deficiency, low vitamin A status and SCI among the groups was compared using Chi-square test. The change in Hb, serum ferritin and body iron store were compared among the four treatment groups using ANOVA test after stratification by SCI. Multiple linear regression was performed to identify the determinants of changes in Hb. In the regression model, baseline concentration of Hb, serum ferritin and transferrin receptor were controlled and treatment effect of iron, vitamin A and their interactions with SCI were included. A general linear model (GLM) with repeated measurements was done to investigate the effect of intervention on the outcome variables such as Hb, iron and vitamin A status and the interaction between the treatment groups and the SCI. The outcome variables were treated as within subject variables. The type of treatment and SCI were treated as between subject variables. The analysis was done with and without including the interaction of treatment and inflammation (treatment × inflammation).

3. Results

Of the 402 recruited subjects, 392 completed the 12 week intervention study but complete data were only available for 391 subjects which was used for the analyses. The main reason for the drop-outs was due to resignation from the schools and data for lost subjects were not different from the remaining subjects. The girls were closely monitored by teachers and the compliance was high (86.5%) with no difference between group. The detail description of baseline characteristics of subjects were described elsewhere [12]. The anthropometric and biochemical status of the subjects were compared among the four groups and there were no significant differences (Table 1).

Table 1. Anthropometric and biochemical status of the subjects at baseline of study [1].

Variables	Treatment Group				Total	p [3]
	FA (n = 98)	VitA (n = 101)	IFA (n = 94)	IFA + VitA (n = 98)	n = 391	
Age (years)	15. 8 ± 1.13	16.1 ± 1.20	15.9 ± 1.21	16.0 ± 1.15	15.9 ± 1.17	0.270
Age of Menarche	13.2 ± 0.89	13.3 ± 0.89	13.1 ± 1.01	13.1 ± 0.89	13.2 ± 0.92	0.469
Weight (kg)	41.3 ± 6.12	42.3 ± 6.42	41.5 ± 6.03	41. 8 ± 6.27	41.7 ± 6.20	0.711
Height (cm)	149.8 ± 4.60	151.7 ± 5.75	150.8 ± 5.33	151.2 ± 5.23	150.8 ± 5.27	0.075
BMI-for-Age	−0.67 ± 0.96	−0.73 ± 0.86	−0.77 ± 0.94	−0.79 ± 1.02	−0.74 ± 0.94	0.821
Thinness (%) [4]	10.2	5.9	9.6	10.2	8.9	0.672
Hemoglobin (g/dL)	8.9 ± 1.3	8.9 ± 1.1	8.8 ± 1.1	8.8 ± 1.2	8.9 ± 1.2	0.850
Serum Ferritin [2] (µg/L)	29.9 ± 2.5	33.4 ± 2.5	31.4 ± 2.4	25.5 ± 3.1	29.9 ± 2.6	0.233
Transferrin receptor (mg/L)	6.74 ± 1.5	6.56 ± 1.42	7.09 ± 1.52	7.16 ± 1.59	6.9 ± 1.5	0.391
Body iron (mg/kg)	4.04 ± 3.77	4.49 ± 4.21	3.97 ± 3.91	3.16 ± 5.27	3.92 ± 4.34	0.185
Serum retinol [2] (µmol/L)	1.18 ± 1.28	1.17 ± 1.31	1.21 ± 1.28	1.18 ± 1.29	1.18 ± 1.29	0.858
Serum folate (nmol/L)	6.02 ± 1.95	6.0 ± 1.7	6.66 ± 1.80	6.27 ± 1.67	6.22 ± 1.78	0.574
CRP(mg/L)	0.73 ± 2.07	0.96 ± 3.32	1.38 ± 6.08	0.47 ± 0.87	0.88 ± 3.60	0.349
AGP (g/L)	0.76 ± 0.18	0.75 ± 0.16	0.74 ± 0.20	0.74 ± 0.17	0.75 ± 0.18	0.798

[1] mean ± SD or geometric mean ± SD, FA: Folate group, VitA: Vitamin A and folate, IFA: Iron and folate group, IFA + vitA: Iron folate and vitamin A group; [2] Serum ferritin and serum retinol were adjusted for sub-clinical inflammation using the meta-analysis correction factor; [3] one-way ANOVA was used for between groups comparison, values are significantly different at $p < 0.05$(Bonferroni correction); [4] Thinness was defined by BMI-for-age z score <−2.

Change in Hb, serum ferritin and body iron store were compared among the four groups after stratification by SCI (Table 2). The results showed no significant difference in Hb change between the 4 treatments although there was a tendency of greater increase in Hb in IFA and IFA + vitA group (p value=0.139) in subjects without SCI, but not in the group with SCI. Regarding the change in serum ferritin and body iron store, there were significant differences between treatment groups in subjects without inflammation and the changes were higher in the IFA and IFA+ VitA and folate groups. However, there were no significant differences between treatments for the changes in Hb, ferritin, or body iron stores in subjects with SCI.

Table 2. Changes in Haemoglobin, serum ferritin and body iron store with or without sub-clinical inflammation at the end of 12 week iron supplementation [1].

	Change in Hb		Change in SF (Log Value)		Change in Body Iron Store	
	No SCI [2] (336)	SCI (55)	No SCI (336)	SCI (55)	No SCI (336)	SCI (55)
FA	1.98 ± 1.08	2.18 ± 1.14	0.91 ± 1.58 [a]	1.05 ± 2.59	−0.24 ± 1.90 [a]	−0.47 ± 1.56
VitA	1.96 ± 1.25	2.12 ± 1.66	1.04 ± 1.65 [a]	0.89 ± 1.87	0.28 ± 1.92 [a]	−0.10 ± 2.22
IFA	2.28 ± 1.18	2.10 ± 1.20	1.26 ± 1.54 [b]	1.05 ± 2.59	1.20 ± 1.64 [b]	0.43 ± 3.23
IFA + vitA	2.28 ± 1.32	2.46 ± 1.40	1.41 ± 1.72 [b]	1.07 ± 1.99	1.73 ± 2.45 [b]	0.78 ± 2.95
F value	1.843	0.186	13.55	0.303	16.477	0.688
p value [3]	0.139	0.906	<0.001	0.823	<0.001	0.563

[1] mean ± SD for Hb change and body iron store, geometric mean ± SD for serum ferritin change, [ab] Values with different superscript letters are significantly different, $p < 0.05$, FA: Folate group, VitA: Vitamin A and folate, IFA: Iron and folate group, IFA + vitA: Iron folate and vitamin A group; [2] SCI = sub-clinical inflammation defined by CRP > 5 mg/L or AGP > 1 g/L at any time throughout 12 week of intervention study; [3] one-way ANOVA was used for between groups comparison, values are significantly different at $p < 0.05$.

Table 3 shows the results of regression analysis for the outcome Hb change in response to iron or vitamin A alone or interaction with inflammation. The result shows treatment with iron significantly increased Hb concentration ($p = 0.002$), but not in the presence of inflammation ($p = 0.268$). Vitamin A alone in those with no inflammation did not increase Hb, but there was a significant increase in Hb in those subjects with inflammation.

Table 3. Determinants for Hb changes from baseline to endline of 12 weeks Intervention.

	Unstandardized Coefficient (B)	SE	Standardized Coefficient	*p*
Treatment with Fe [1]				
(Constant)	11.19	0.51		<0.001
Hb at baseline	−0.85	0.04	−0.81	<0.001
Serum ferritin at baseline	−0.09	0.11	−0.03	0.397
sTfR at BL	−1.85	0.27	−0.27	<0.001
Treatment with Iron	0.28	0.09	0.11	0.002
SCI treated with Fe	−0.26	0.24	−0.05	0.268
SCI	0.11	0.16	0.03	0.490
Treatment with VitA [2]				
(Constant)	11.37	0.51		<0.001
Hb at baseline	−0.85	0.04	−0.82	<0.001
Serum ferritin at baseline	−0.11	0.11	−0.04	0.324
sTfR at BL	−1.77	0.27	−0.25	0.000
Treatment with VitA	−0.08	0.09	−0.03	0.359
SCI treated with VitA	0.48	0.24	0.10	0.044
SCI	−0.26	0.17	−0.07	0.123

[1] Linear regression for the change in haemoglobin concentration by treatment with Iron, adjusted for baseline Hb, baseline serum ferritin and sTfR, $R^2 = 0.57$, F(89.34), ($p < 0.001$); [2] Linear regression for the change in haemoglobin concentration by treatment with Vitamin A, adjusted for baseline Hb, baseline serum ferritin and sTfR, $R^2 = 0.57$, F(86.97), ($p < 0.001$).

The results of analysis by general linear model with repeated measures for subjects with and without inflammation are shown in Table 4. There were significant differences in the increases in Hb ($p < 0.027$) and sTfR ($p < 0.013$) concentrations in response to the different treatments between those subjects with and without inflammation. The results showed that in the group with SCI, subjects who received either vitamin A alone or together with IFA had better iron status compared to the other groups.

Table 4. Effect of iron and vitamin A supplementation in the presence and absence of sub-clinical inflammation among the adolescent girls during 12 weeks of intervention [1].

Variables	Without Sub-Clinical Inflammation				With Sub-Clinical Inflammation				p [2]			
	FA (n = 82)	VitA (n = 86)	IFA (n = 82)	IFA + vitA (n = 86)	FA (n = 16)	VitA (n = 15)	IFA (n = 12)	IFA + vitA (n = 12)	Within Group Over Time	Between Group	SCI	Group × SCI
Hb(g/L)												
Baseline	9.0 ± 1.3	8.9 ± 1.1	8.9 ± 1.1	8.8 ± 1.2	8.6 ± 1.3	9.1 ± 1.4	8.3 ± 1.1	8.9 ± 1.2	<0.001	0.181	0.325	0.027
Midline	9.9 ± 1.1	9.8 ± 0.9	9.8 ± 1.0	9.9 ± 1.0	9.7 ± 1.0	10.4 ± 1.1	9.7 ± 1.6	9.8 ± 1.2				
Endline	11 ± 0.8	10.9 ± 0.9	11.2 ± 0.9	11.1 ± 0.9	10.7 ± 0.8 [b]	11.2 ± 1.1 [b]	10.4 ± 1 [a]	11.3 ± 0.9 [b]				
SF (µg/L) [3]												
Baseline	43.7 ± 46.2	43.4 ± 33.3	40.7 ± 28.0	36.9 ± 28.9	54.5 ± 46.7	79.8 ± 59.9	67.7 ± 51.9	65.9 ± 52	0.178	0.154	<0.002	0.444
Midline	34.2 ± 32.0	39.5 ± 29.6	39.8 ± 22.9	38.1 ± 25.9	46.0 ± 40.1	65.5 ± 38.6	51.4 ± 27.9	56.5 ± 28.4				
Endline	38.2 ± 34.8	44.4 ± 31.3	47.3 ± 29.1	45.4 ± 30.6	50.7 ± 44.6	68.9 ± 42	68.2 ± 48.7	59 ± 42				
sTfR(mg/L)												
Baseline	7.3 ± 5.8	7.2 ± 3.1	7.5 ± 4.5	8.3 ± 5.7	8.8 ± 4.7	6.3 ± 1.6	11.2 ± 9	7.5 ± 4.5	<0.001	0.039	0.129	0.013
Midline	7.1 ± 5.7	7.0 ± 3.3	7.0 ± 4.4	7.7 ± 5.0	9.1 ± 6.1 [b]	5.9 ± 1.4 [a]	11.0 ± 8.3 [b]	6.6 ± 2.9 [a]				
Endline	7.2 ± 5.5	6.9 ± 3.2	6.8 ± 4.2	6.9 ± 3.8	9.0 ± 4.9 [b]	5.7 ± 1.3 [a]	10.5 ± 8.2 [b]	6.1 ± 2.0 [a]				
Body Iron Store (mg/kg)												
Baseline	4.1 ± 3.5	4.1 ± 4.3	3.9 ± 3.7	2.9 ± 5.2	3.8 ± 5.3	6.9 ± 2.8	4.4 ± 5.1	5.1 ± 5.5	0.003	0.110	0.054	0.128
Midline	3.5 ± 3.3	3.9 ± 4.0	4.5 ± 2.9	3.9 ± 3.7	3.6 ± 3.9 [a]	6.8 ± 2.3 [b]	4.0 ± 4.1 [a]	6.0 ± 3 [b]				
Endline	3.9 ± 3.7	4.4 ± 4.2	5.1 ± 3.1	4.6 ± 4.0	3.3 ± 5.4	6.8 ± 3.1	4.8 ± 4.6	5.9 ± 3.7				
Serum Retinol (µmol/L) [3]												
Baseline	1.23 ± 0.36	1.22 ± 0.32	1.25 ± 0.31	1.19 ± 0.31	1.08 ± 0.22	1.1 ± 0.29	1.14 ± 0.27	1.32 ± 0.33	0.052	0.329	0.338	0.80
Midline	1.12 ± 0.31	1.19 ± 0.35	1.18 ± 0.29	1.18 ± 0.32	1.08 ± 0.3	1.17 ± 0.34	1.28 ± 0.31	1.15 ± 0.27				
Endline	1.11 ± 0.32	1.23 ± 0.35	1.18 ± 0.34	1.19 ± 0.34	1.03 ± 0.16	1.13 ± 0.23	1.1 ± 0.32	1.16 ± 0.28				

[1] Values were presented as mean ± SD. [ab] Values with different superscript letters are significantly different, $p < 0.05$, FA: Folate group, VitA: Vitamin A and folate, IFA: Iron and folate group, IFA + VitA: Iron folate and vitamin A group; [2] Repeated measures ANOVA with the test for 2-factor interaction, significant at p < 005; [3] Transformed data for SF, sTfR, and Serum Retinol were used for analyses.

4. Discussion

The important finding of the study is that the effect of iron supplementation can be hampered by SCI and the addition of vitamin A can be an advantage in that condition.

The study was conducted in the delta region Myanmar where the prevalence of anemia was reportedly high [24]. Indeed, the finding from the screening phase showed the prevalence of anemia was ~ 60% in the area which confirmed anemia as a major public health problem in the area [13]. Myanmar is a country situated in South East Asia with the population of approximately 51 millions according to census 2014 [25]. About 70% of the populations reside in rural area and majority of them are working in the agricultural sector. The results showed that around 30% of the subjects were iron deficient (serum ferritin <15 μg/L and/or sTfR >8.5 mg.L) and had low vitamin A status (serum retinol <1.05 μmol/L) (Table 1). The prevalence of thinness (BMI-for-age < e-2) was 8.9% among the girls which was lower than findings from recently conducted nationwide micronutrient survey (18% prevalence of thinness for union level) [26].The subjects from the study area were relatively healthy and as a result, the prevalence of inflammation was as low as 6%. This was in contrast with our previous work from Indonesia where the prevalence of inflammation was almost 40% among high school girls of the same age group [27].

However, despite the low prevalence, the finding showed the significant negative impact of inflammation on effectiveness of iron supplementation. Table 2 shows the influence of SCI on the effect of iron supplementation for the changes in Hb, serum ferritin and body iron at the end of the 12 week supplementation. The finding suggested that effect of iron supplementation on serum ferritin and body iron store were impaired by SCI. Inflammation is our body physiological response to infections and injuries and the purpose is to minimize the damage to our body and to facilitate the repair process [28]. Unlike in clinical settings, sub-clinical inflammation (SCI) can be always present among the apparently healthy population and unless taken into account, it will lead to misinterpretation of the biomarkers of nutritional status assessment [9,10]. Few studies have investigated the association between sub-clinical inflammation and anemia, but were mostly inconclusive and descriptive. A recently published paper showed subclinical inflammation (AGP>1.0 g/L) was a risk factor for anemia among the Cambodian children and the author suggested that there might be inhibition of iron absorption as well as the release of stored iron as a result of inflammation [29]. Another study from Papua New Guinea showed the children who had high CRP (>5 mg/L) and AGP (>1.2 mg/L) had high prevalence of anemia [30]. While it failed to reach statistical significance ($p = 0.053$), a study from Nicaragua showed the risk of anemia was 1.5 times higher for the subjects with high acute phase protein (AGP > 1.0 g/L) than the normal subjects [31]. Similarly, the finding from our study indicates that SCI is an important determinant that undermines the effectiveness of iron supplementation.

The mechanism by which inflammation interferes with the iron supplementation might be explained by the negative iron regulatory hormone hepdicin, which inhibits intestinal iron absorption, release of storage iron from the reticuloendothelial system. During inflammation, hepcidin is released mainly from the liver with the stimulation from pro-inflammatory cytokines especially IL-6 [7,32]. It is also suggested that proinflmammatory cytokines suppress the production or biological activity of erythropoietin in anemia of inflammation [33]. However, studies have not been done to investigate the influence of SCI on iron supplementation and despite the low prevalence of SCI, we have shown that SCI interfered with the effectiveness of iron supplementation.

Vitamin A has been extensively studied for its important roles in body's physiological and immune functions yet still many issues remain unsolved and the data is limited especially for this age group. The combined effect of vitamin A and iron supplementation was well documented in previous studies and studies showed that vitamin A is essential for the mobilization of storage iron and a possible role in enhancing the erythropoietin [34,35]. The regression analysis for the outcome Hb change shows the interaction effect between vitamin A and SCI (vitamin A × SCI) (Table 3). Although we cannot conclude with certainty, the possible mechanism might be the immunomodulatory function of vitamin

A whereas it enhances Th2 anti-inflammatory pathway while minimizing the Th1 pro-inflammatory pathway [36]. Moreover, this effect may be an adjuvant function of vitamin A on iron supplementation.

Table 4 shows the effect of iron supplementation is hampered by SCI. The important finding is the significant interaction effect between the treatment and SCI (treatment × SCI) for the outcome hemoglobin and serum transferrin receptor. Specifically, when there was SCI, those who received vitamin A such as vitamin A group and vitamin A + IFA group showed significantly better outcome for Hb and sTfR compared to the other groups. In this study, both hemoglobin and serum transferrin receptor concentration were not influenced by SCI but when accompanied by SCI, vitamin A containing group showed better responses for these indicators. There was no interaction between iron and SCI. On the other hand, when there was no SCI, IFA, and vitamin A +IFA group showed better outcomes. These findings suggested that iron supplementation would be more effective regardless of vitamin A if there is no inflammation. However, once there is inflammation, the additional vitamin A supplementation is an advantage.

The significance of the study is that the effect of iron supplementation on iron status can be observed in the two scenarios stratified by the inflammation status. The prevalence of SCI was relatively low in this study, but there might be seasonal increases that should be further explored. The results of this study indicate that SCI has programmatic implications for the effectiveness of iron supplementation where infections are common. Given SCI is not commonly assessed in programmatic setting, additional vitamin A may be an advantage to improve the effectiveness of IFA supplementation.

Author Contributions: M.K.H., D.D., A.A., B.U., U.F. and D.I.T. designed the research; M.K.H. conducted the research; M.K.H., U.F. and D.I.T. analyzed the data; M.K.H., U.F. and D.I.T. wrote the paper and M.K.H. had primary responsibility for the final content. All authors read and approved the final manuscript.

Funding: The study was funded by a DAAD (German Academic Exchange Service) scholarship and the Nestle' Foundation. DAAD and Nestle' Foundation had no influence in the design or analysis of the study, writing of this article or the decision to submit it for publication. This publication was funded by Directorate of Research and Community Engagement of Universitas Indonesia (Number: 5282/UN2.R3.1/HKP05.00/2018).

Acknowledgments: The authors thank Director General Myo Khin from Department of Medical Research, Lower Myanmar, Theingi Thwin, Deputy Director and team from Nutrition Research Division, Department of Medical Research, Lower Myanmar for biochemical data collection, J.G. Erhardt for for analyses of RBP, serum ferritin, sTfR, CRP and AGP and for his kind inputs and discussions.

Conflicts of Interest: The authors declare no conflict of interest.

References

1. McLean, E.; Cogswell, M.; Egli, I.; Wojdyla, D.; De Benoist, B. Worldwide prevalence of anaemia, WHO Vitamin and Mineral Nutrition Information System, 1993–2005. *Public Health Nutr.* **2008**, *12*, 444–454. [CrossRef]
2. Zimmermann, M.B.; Hurrell, R.F. Nutritional iron deficiency. *Lancet* **2007**, *370*, 511–520. [CrossRef]
3. Grimm, K.A.; Sullivan, K.M.; Alasfoor, D.; Parvanta, I.; Suleiman, A.J.M.; Kaur, M.; Al-Hatmi, F.O.; Ruth, L.J. Iron-fortified wheat flour and iron deficiency among women. *Food Nutr. Bull.* **2012**, *33*, 180–185. [CrossRef] [PubMed]
4. Johnson, E.E.; Wessling-Resnick, M. Iron metabolism and the innate immune response to infection. *Microbes Infect.* **2012**, *14*, 207–216. [CrossRef] [PubMed]
5. Sazawal, S.; Black, R.E.; Ramsan, M.; Chwaya, H.M.; Stoltzfus, R.J.; Dutta, A.; Dhingra, U.; Kabole, I.; Deb, S.; Othman, M.K.; et al. Effects of routine prophylactic supplementation with iron and folic acid on admission to hospital and mortality in preschool children in a high malaria transmission setting: Community-based, randomised, placebo-controlled trial. *Lancet* **2006**, *367*, 133–143. [CrossRef]
6. Brand-Miller, J.C.; Liu, V.; Petocz, P.; Baxter, R.C. The glycemic index of foods influences postprandial insulin-like growth factor–binding protein responses in lean young subjects. *Am. J. Clin. Nutr.* **2005**, *82*, 350–354. [CrossRef]
7. Ganz, T.; Nemeth, E. Hepcidin and iron homeostasis. *Biochim. Biophys. Acta (BBA) Mol. Cell Res.* **2012**, *1823*, 1434–1443. [CrossRef] [PubMed]

8. Thurnham, D.I.; Mburu, A.S.; Mwaniki, D.L.; De Wagt, A. Micronutrients in childhood and the influence of subclinical inflammation. *Proc. Nutr. Soc.* **2005**, *64*, 502–509. [CrossRef] [PubMed]
9. Thurnham, D.I.; McCabe, G.P.; Northrop-Clewes, C.A.; Nestel, P. Effects of subclinical infection on plasma retinol concentrations and assessment of prevalence of vitamin A deficiency: Meta-analysis. *Lancet* **2003**, *362*, 2052–2058. [CrossRef]
10. Thurnham, D.I.; McCabe, L.D.; Haldar, S.; Wieringa, F.T.; Northrop-Clewes, C.A.; McCabe, G.P. Adjusting plasma ferritin concentrations to remove the effects of subclinical inflammation in the assessment of iron deficiency: A meta-analysis. *Am. J. Clin. Nutr.* **2010**, *92*, 546–555. [CrossRef]
11. Northrop-Clewes, C.A. Interpreting indicators of iron status during an acute phase response—Lessons from malaria and human immunodeficiency virus. *Ann. Clin. Biochem.* **2008**, *45*, 18–32. [CrossRef] [PubMed]
12. Htet, M.K.; Fahmida, U.; Dillon, D.; Akib, A.; Utomo, B.; Thurnham, D.I. The influence of vitamin A status on iron-deficiency anaemia in anaemic adolescent schoolgirls in Myanmar. *Public Health Nutr.* **2013**, *17*, 2325–2332. [CrossRef] [PubMed]
13. Htet, M.K.; Dillon, D.; Akib, A.; Utomo, B.; Fahmida, U.; Thurnham, D.I. Microcytic anaemia predominates in adolescent school girls in the delta region of Myanmar. *Asia Pac. J. Clin. Nutr.* **2012**, *21*, 411–415. [PubMed]
14. Angeles-Agdeppa, I.; Schultink, W.; Sastroamidjojo, S.; Gross, R.; Karyadi, D. Weekly micronutrient supplementation to build iron stores in female Indonesian adolescents. *Am. J. Clin. Nutr.* **1997**, *66*, 177–183. [CrossRef] [PubMed]
15. WHO. *Weekly Iron and Folic Acid Supplementation Programmes for Women of Reproductive Age: An Analysis of Best Programme Practices*; World Health Organization: Geneva, Switzerland, 2011.
16. WHO. *WHO AnthroPlus for Personal Computers Manual: Software for Assessing Growth of the World's Children and Adolescents*; World Health Organization: Geneva, Switzerland, 2009.
17. Erhardt, J.G.; Estes, J.E.; Pfeiffer, C.M.; Biesalski, H.K.; Craft, N.E. Combined Measurement of Ferritin, Soluble Transferrin Receptor, Retinol Binding Protein, and C-Reactive Protein by an Inexpensive, Sensitive, and Simple Sandwich Enzyme-Linked Immunosorbent Assay Technique. *J. Nutr.* **2004**, *134*, 3127–3132. [CrossRef] [PubMed]
18. Cook, J.D.; Flowers, C.H.; Skikne, B.S. The quantitative assessment of body iron. *Blood* **2003**, *101*, 3359–3363. [CrossRef]
19. Erhardt, J.G.; Mack, H.; Sobeck, U.; Biesalski, H.K. β-Carotene and α-tocopherol concentration and antioxidant status in buccal mucosal cells and plasma after oral supplementation. *Br. J. Nutr.* **2002**, *87*, 471–475. [CrossRef]
20. O'Broin, S.; Kelleher, B. Microbiological assay on microtitre plates of folate in serum and red cells. *J. Clin. Pathol.* **1992**, *45*, 344–347. [CrossRef]
21. Gibson, R. *Principles of Nutritional Assessment*; Oxford University Press: New York, NY, USA, 2005.
22. WHO. *Iron Deficiency Anaemia: Assessment, Prevention, and Control. A Guide for Programme Managers*; World Health Organization: Geneva, Switzerland, 2001.
23. West, K.P. Extent of vitamin A deficiency among preschool children and women of reproductive age. *J. Nutr.* **2002**, *132*, S2857–S2866. [CrossRef] [PubMed]
24. National Nutrition Centre (NNC). *National Haemoglobin and Nutritional Status Survey among Adolsecents*; Department of Health, Ministry of Health: Yangon, Myanmar, 2002.
25. Department of Population. *The 2014 Myanmar Population and Housing Census*; Ministry of Immigration and Population: Nay Pyi Taw, Myanmar, 2014.
26. National Nutrition Centre (NNC). *Myanmar Micronutrients and Food Consumption Survey (2017–2018), Interim Report*; Department of Health, Ministry of Health and Sports: Nay Pyi Taw, Myanmar, 2019.
27. Htet, M.K.; Dillon, D.; Rosida, A.; Timan, I.; Fahmida, U.; Thurnham, D.I. Hepcidin profile of anemic adolescent schoolgirls in Indonesia at the end of 12 weeks of iron supplementation. *Food Nutr. Bull.* **2014**, *35*, 160–166. [CrossRef]
28. Steel, D.M.; Whitehead, A.S. The major acute phase reactants: C-reactive protein, serum amyloid P component and serum amyloid A protein. *Immunol. Today* **1994**, *15*, 81–88. [CrossRef]
29. George, J.; Yiannakis, M.; Main, B.; Devenish, R.; Anderson, C.; An, U.S.; Williams, S.M.; Gibson, R.S. Genetic Hemoglobin Disorders, Infection, and Deficiencies of Iron and Vitamin A Determine Anemia in Young Cambodian Children. *J. Nutr.* **2012**, *142*, 781–787. [CrossRef] [PubMed]
30. Baumgartner, J.; Smuts, C.M.; Malan, L.; Arnold, M.; Yee, B.K.; Bianco, L.E.; Boekschoten, M.V.; Müller, M.; Langhans, W.; Hurrell, R.F.; et al. In Male Rats with Concurrent Iron and (n-3) Fatty Acid Deficiency,

Provision of Either Iron or (n-3) Fatty Acids Alone Alters Monoamine Metabolism and Exacerbates the Cognitive Deficits Associated with Combined Deficiency. *J. Nutr.* **2012**, *142*, 1472–1478. [CrossRef] [PubMed]

31. Sullivan, K.M.; Venugopalan, B.; Jefferds, M.E.; Boy, E.; Bonilla, J.; Sandino, I.; Hallesleven, P. Association of elevated α_1-acid glycoprotein (AGP) and the prevalence of anemia in Nicaraguan preschool children. *Food Nutr. Bull.* **2012**, *33*, 137–141. [CrossRef] [PubMed]
32. Nemeth, E.; Rivera, S.; Gabayan, V.; Keller, C.; Taudorf, S.; Pedersen, B.K.; Ganz, T. IL-6 mediates hypoferremia of inflammation by inducing the synthesis of the iron regulatory hormone hepcidin. *J. Clin. Investig.* **2004**, *113*, 1271–1276. [CrossRef]
33. Dobrian, A.D.; Lieb, D.C.; Cole, B.K.; Taylor-Fishwick, D.A.; Chakrabarti, S.K.; Nadler, J.L. Functional and pathological roles of the 12- and 15-lipoxygenases. *Prog. Lipid Res.* **2011**, *50*, 115–131. [CrossRef]
34. Zimmermann, M.B.; Biebinger, R.; Rohner, F.; Dib, A.; Zeder, C.; Hurrell, R.F.; Chaouki, N. Vitamin A supplementation in children with poor vitamin A and iron status increases erythropoietin and hemoglobin concentrations without changing total body iron. *Am. J. Clin. Nutr.* **2006**, *84*, 580–586. [CrossRef]
35. Semba, R.; Bloem, M. The anemia of vitamin A deficiency: Epidemiology and pathogenesis. *Eur. J. Clin. Nutr.* **2002**, *56*, 271–281. [CrossRef]
36. Villamor, E.; Fawzi, W.W. Effects of Vitamin A Supplementation on Immune Responses and Correlation with Clinical Outcomes. *Clin. Microbiol. Rev.* **2005**, *18*, 446–464. [CrossRef]

nutrients

MDPI

Article

The Association of *TMPRSS6* Gene Polymorphism and Iron Intake with Iron Status among Under-Two-Year-Old Children in Lombok, Indonesia

Dewi Shinta [1,2], Asmarinah [3], Chris Adhiyanto [4], Min Kyaw Htet [1,5] and Umi Fahmida [1,*]

1 Southeast Asian Ministers of Education Organization Regional Center for Food and Nutrition (SEAMEO RECFON) Pusat Kajian Gizi Regional Universitas Indonesia, Jakarta 10430, Indonesia; dewishinta.deos@gmail.com (D.S.); kyawhtet@gmail.com (M.K.H.)

2 Study Program in Nutrition, Faculty of Medicine, Universitas Indonesia, Jakarta 10430, Indonesia

3 Department of Medical Biology, Faculty of Medicine, Universitas Indonesia, Jakarta 10430, Indonesia; asmarinah.si@gmx.de

4 Faculty of Medicine, Universitas Islam Negeri Syarif Hidayatullah, Jakarta 15412, Indonesia; chrisbiomed@uinjkt.ac.id

5 Sydney School of Public Health, Sydney Medical School, The University of Sydney, Sydney, NSW 2006, Australia

* Correspondence: umifahmida@gmail.com; Tel.: +62-21-31930205; Fax: +62-21-3913933

Received: 15 February 2019; Accepted: 29 March 2019; Published: 19 April 2019

Abstract: Multiple common variants in *transmembrane protease serine 6 (TMPRSS6)* were associated with the plasma iron concentration in genome-wide association studies, but their effect in young children where anemia and iron deficiency (ID) were prevalent has not been reported, particularly taking account of iron intake. This study aims to investigate whether *TMPRSS6* SNPs (rs855791 and rs4820268) and iron intake are associated with a low iron and hemoglobin concentration in under-two-year-old children. The study analyzed the baseline of a randomized trial (NUPICO, ClinicalTrials.gov NCT01504633) in East Lombok, Indonesia. Children aged 6–17 months ($n = 121$) were included in this study. The multiple linear regressions showed that *TMPRSS6* decreased serum ferritin (SF) by 4.50 g/L per copy minor allele (A) of rs855791 ($p = 0.08$) and by 5.00 μg/L per copy minor allele (G) of rs4820268 ($p = 0.044$). There were no associations between rs855791 and rs4820268 with soluble transferrin receptor (sTfR) and hemoglobin (Hb) concentration (rs855791; $p = 0.38$ and $p = 0.13$, rs4820268; $p = 0.17$ and $p = 0.33$). The finding suggests the need for further studies to explore whether the nutrient recommendation for iron should be based on genetic characteristics, particularly for children who have mutation in *TMPRSS6*.

Keywords: anemia; iron deficiency; *TMPRSS6*; iron intake; children; Indonesia

1. Introduction

Worldwide, Iron Deficiency Anemia (IDA) affects 24.8% of the total population and the highest prevalence is among under-five-year-old children (47.4%) [1], particularly in their first two years of life. Studies often reported a decline in the nutritional status during the first two years of life and therefore this period is a critical window of opportunity for improving childhood nutrition [2].

Previously, IDA has been known to be associated with dietary and/or other environmental factors such as infection. Recently, several studies have indicated a genetic contribution in the development of ID, i.e., 20–30% of variability in iron concentration is attributable to genetic factors [3]. Genome-wide association (GWA) studies have revealed a number of genetic variants of human and animal genomes which influence iron concentration. Studies in the mutant mouse showed that the *transmembrane protease serrin 6 (TMPRSS6)* genes caused the loss of the catalytic domain of matriptase-2 that concurrently increased the hepcidin level in the liver. Increased hepcidin level led to a severe microcytic anemia due to low iron absorption in the intestine [4]. Findings of severe IDA in masked mice have led to further investigations on the *TMPRSS6* mutations in humans with iron refractory iron deficiency anemia (IRIDA) [5].

Recently, GWA studies identified several single nucleotide polymorphisms (SNPs) of the *TMPRSS6* genes that influence microcytic red blood cell phenotypes. Several SNPs of *TMPRSS6* have been identified to be associated with IDA. Among these SNPs, rs855791 and rs4820268 have the strongest association either on red blood cell indices or iron parameters especially in the Asian population [5]. No previous studies have reported the influence of the *TMPRSS6* gene on iron status indicators in under-two-year-old children. In addition, most studies reporting the influence of genetic factors on the iron status did not take into account iron intake at the same time. Therefore, this study aims to investigate whether minor alleles of the *TMPRSS6* gene at rs855791 and rs4820268 as well as the usual iron intake are associated with low iron and hemoglobin concentrations among under-two-year-old children in East Lombok, where the prevalence of anemia and iron deficiency was reported to be high.

2. Material and Method

2.1. Study Population

Data for this study were collected at the baseline of a randomized trial of NUPICO (Can Nutrigenetics help explain the Mixed Result on Effect of LCPUFAs and Iron on Child Cognition?) that aims to assess gene-nutrient interaction in explaining effect of LC-PUFAs and iron on cognitive functioning of young children. The NUPICO was registered at ClinicalTrials.gov as NCT01504633. Ethical approval was obtained from the Ethical Commission of Faculty of Medicine, Universitas Indonesia (No. 586/PT.02.FK/ETIK/2011, with addendum No.837/UN2.F1/ETIK/IX/2015). Written informed consent was obtained from the parents or caregivers of the children. The study was conducted in 2012 in the East Lombok district, West Nusa Tenggara province, Indonesia. Breastfed, 12–17 months old children of *Sasak* ethnicity were recruited with the following exclusion criteria: Low birth weight (<2500 g), having congenital disorder, severe anemia (hemoglobin, Hb < 70 g/L) and malaria infection. All subjects with complete data on genotype, iron indicators, and iron intake were included in this analysis ($n = 121$). This sample size met the minimum sample ($n = 94$) which was calculated based on the coefficient correlation of iron intake and Hb ($r = 0.233$, $p < 0.05$) [6] with 0.05 significance level and 80% power.

2.2. Dietary Iron Intake Assessment

The usual iron intake of the children was assessed using a validated semi-quantitative food frequency questionnaire (SQ-FFQ) of the previous month intake. The SQ-FFQ represented 90% of iron intake of the study population and was validated using two-day non-consecutive 24 h dietary recalls and showed good correlation ($p < 0.05$) and no significant mean difference in the estimated usual iron intake. The proportion at risk of inadequate iron intake was estimated from two-day non-consecutive 24 h dietary recalls using the IMAPP Version 1.0.

2.3. Iron and Hemoglobin Measurements

The concentration of Iron was assessed by a soluble serum transferrin receptor (sTfR) and serum ferritin (SF), which were analyzed using the Immunosorbent assays (ELISA) method. To correct the influence of subclinical inflammation which was defined by the C-reactive protein (CRP) (>5 mg/dL) and α-1 Acid Glycoprotein (AGP) (>1 g/dL), correction factors were used to adjust the SF concentration following the Thurnham model, i.e., 1.0 (healthy stage), 0.77 (incubation stage, raised CRP only), 0.53 (acute inflammation, raised CRP and AGP), 0.75 (chronic inflammation, raised AGP only) [7]. The concentration of Hb was analyzed using the cyanmethemoglobin method. The body iron store was calculated using SF and sTfR following the method by Cook et al. (body iron mg/kg = −[log(R/F ratio)−2.8229]/0 [8]. Anemia was defined by Hb < 11 g/dL, ID by sTfR concentration > 8.5 mg/L, and/or SF < 12 μg/mL, whereas IDA was defined by a combination of both anemia and ID.

2.4. Genotyping Procedure

DNA extraction was done using the Gene Aid Kit (#GB300, Taiwan, Republic of China). The SNP genotyping analysis was done to identify the Allele type of rs855791 and rs4820268 as the *TMPRSS6* gene variants. The method used Taqman-assay with two allele-specific probes and a pair of primers to specifically detect the type of genotype/allele, i.e., the Applied Biosystem TaqMan SNP Genotyping Assay ID C_11885329_10 for rs855791, ID C_32899902_10 for rs4820268 and GTX press Master mix kits (Cat. 94404, New York, NY, USA) kits. The Step One^{TM} Real Time PCR-system with a 48-well thermo bloc instrument and the stepOneV2.3 software (AB) was used for the analysis of SNPs genotyping assay. A representative sample (twelve replicates) was repeated to confirm the genotype and the success rate was 100%.

2.5. Statistical Analysis

Statistical analyses were performed using the SPSS statistical software for Windows (version 20.0; SPSS, Inc., Chicago, IL, USA). Continuous variables (iron intake, sTfR, SF and Hb) were tested for its normal distribution using the one-sample Kolmogorov-Smirnov test. Distributions of continuous variables were presented as mean ± SD (for normally distributed data) and median (25th to 75th) (for non-normally distributed data). Polymorphism of the *TMPRSS6* gene, SNP rs855791 and rs4820268 were presented as a percentage. Multiple linear regression analyses were conducted separately between SNP rs855791 and rs4820268 because of existing multicollinearity between the two SNPs, as indicated by the value of variants inflation factor (VIF) > 1/(1-R^2) and tolerance (TOL) < (1-R^2).

3. Results

Table 1 showed that the homozygote minor allele frequency (MAF) genotype (AA) of rs855791 and (GG) of rs4820268 were the biggest proportion of their genotype profile. The Hardy-Weinberg Equilibrium (HWE) test showed that the (q) variant allele frequency of rs855791 (A) and rs4820268 (G) were 74% and 71%, respectively. Whereas the (p) wild-type allele frequency of rs855791 (G) and rs4820268 (A) were 26% and 29%. Two-thirds (67.7%) of the children were also found to be at risk of having inadequate iron intake.

Table 1. Characteristics of the study population (N = 121).

Factors	*n* (%)
Demographic	
Age of children * [month]	14.12 ± 1.39
Sex of children	
Boy	58 (47.9)
Girl	63 (52.1)
Maternal Education	
Primary school	81 (67.0)
Secondary school	40 (37.0)
Maternal Occupation	
Not working	88 (72.7)
Working	33 (27.3)

Table 1. *Cont.*

Factors	*n* (%)
Genotype Distribution	
Polymorphism of rs855791	
GG homozygote	10 (8.30)
GA heterozygote	43 (35.50)
AA homozygote	68 (56.20)
Polymorphism of rs4820268	
AA homozygote	12 (9.90)
AG heterozygote	45 (37.20)
GG heterozygote	64 (52.90)
Minor Allele Frequency (MAF)	
SNP rs855791 [A/G]	179 (73.90)
SNP rs4820268 [G/A]	173 (71.40)
Intake	
Iron [mg/day] *	4.5 ± 3.01
Iron indicator	
Serum ferritin [μg/L] *,†	14.7 ± 18.00
Serum transferrin receptor [mg/L] *	8.7 ± 3.03
Hemoglobin [g/dL] *	9.51 ± 1.03
Body iron store [mg/kg] *	−0.9 ± 3.8
Anemia	112 (92.6)
Iron deficiency	84 (69.4)
Iron Deficiency Anemia	77 (63.6)

* The data was presented as mean ± SD. † Serum ferritin was adjusted by subclinical inflammation using Thurnham method as specified in the method section.

Based on data analysis from a total of 121 children, 69.4% suffered from iron deficiency, 92.6% had anemia and 63.6% had iron deficiency anemia. We found an increasing trend of iron deficiency, anemia and iron deficiency anemia in children who had minor allele (A) as iron lowering allele (ILA) in rs855791 and minor allele (G) in rs4820268 (Figures 1 and 2).

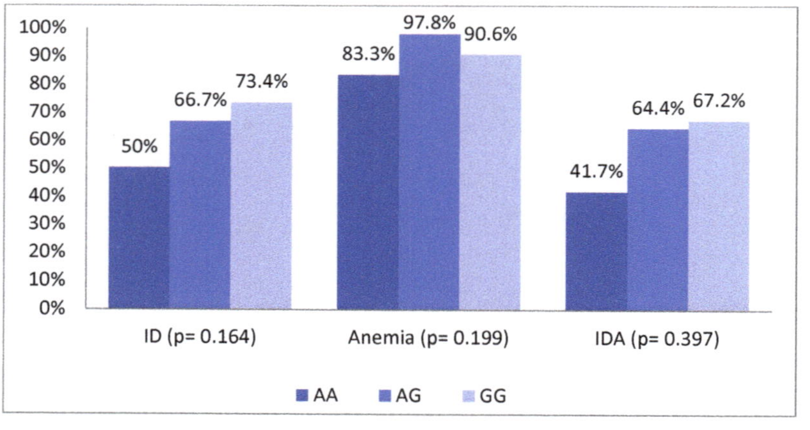

Figure 1. Distribution of Iron deficiency, anemia and iron deficiency anemia status across genotypes of SNPs rs4820268.

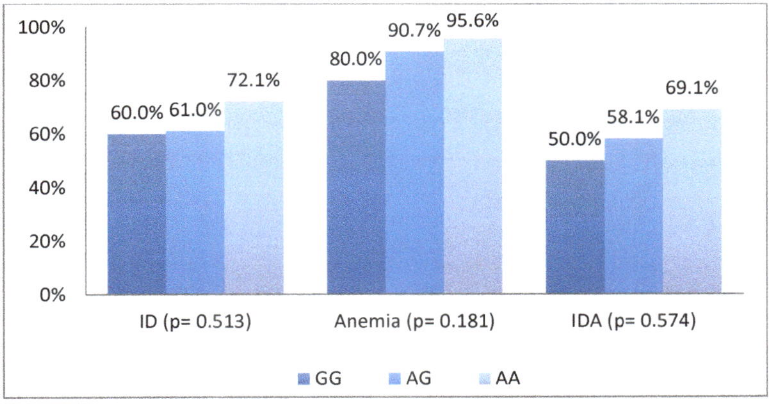

Figure 2. Distribution of Iron deficiency, anemia and iron deficiency anemia status across genotypes of SNPs rs855791.

There were comparable characteristics in terms of sex, age, and the usual iron intake from complementary feeding between genotypes of rs855791 and rs4820268, with the exception of AGP by genotype of rs4820268 ($p < 0.05$), in which AGP is lower with more (G) allele in this SNP (Table 2).

Table 2. Characteristics and iron status of the subjects, stratified by genotypes of SNPs rs855791 and rs420268.

	SNP rs855791				SNP rs420268			
	GG (n = 10)	GA (n = 43)	AA (n = 68)	p	AA (n = 12)	AG (n = 45)	GG (n = 64)	p
Age [mo]	14.08 ± 1.4	14.19 ± 1.3	14.1 ± 1.5	0.945	14.3 ±1.3	14.1 ± 1.2	14.1 ± 1.5	0.875
Sex [boy] *	6 (5)	18 (14.9)	34 (28.1)	0.823	5 (4.1)	21 (17.4)	32 (26.4)	0.812
Iron intake [mg]	4.26 ± 3.17	4.7 ±3.62	4.49 ± 2.57	0.877	3.8 ± 2.9	4.7 ± 3.5	4.5 ± 2.5	0.638
CRP [mg/L]	5.7 ± 7.3	4.4 ± 8.4	3.4 ± 5.8	0.568	6.7 ± 10.1	3.9 ± 7.2	3.5 ± 5.9	0.351
AGP [mg/L]	1 ± 0.3	0.9 ± 0.3	0.8 ± 0.2	0.072	0.9 ± 0.1	0.8 ± 0.3	0.7 ± 0.2	0.049
Serum ferritin [μg/L] †	11.4 ± 7.7	13.9 ± 14.9	10.5 ± 9.7	0.314	12.1 ± 8	14.1 ± 6	10 ± 7.7	0.197
Serum transferrin receptor [mg/L]	7.4 ± 2.5	8.8 ± 2.9	8.8 ± 3.1	0.839	7.4 ± 2.3	8.7 ± 2.8	8.9 ± 3.2	0.290
Body iron store [mg/Kg]	0.5 ± 3.5	−0.4 ± 4.0	−1.4 ± 3.8	0.180	−1.5 ± 3.6	−0.4 ± 4.1	−0.7 ± 3.5	0.123
Hemoglobin [g/dL]	10.04 ± 1.07	9.53 ± 1.08	9.41 ± 0.97	0.195	10.08 ± 1.05	9.37 ± 0.90	9.50 ± 1.08	0.103

* Sex (boy) was presented in n (%), † Serum ferritin (SF) was adjusted by. subclinical inflammation using Thurnham method as specified in the method section.

The multiple linear regressions showed that after controlling for the usual iron intake, sex and age, per copy minor allele (G) was significantly associated with lower SF by 5.00 μg/L of rs4820268 (p = 0.044) and borderline significantly associated with lower SF by 4.50 μg/L of rs855791 (p = 0.080) (Table 3).

Table 3. Association between selected SNPs of *TMPRSS6* with Serum Ferritin (SF), Soluble Transferrin Receptor (sTfR), Hemoglobin (Hb) and body iron store.

Factors	Serum Ferritin (µg/L) *			Serum Transferrin Receptor (mg/L)			Hemoglobin (g/dL)			Body Iron Store (mg/Kg Body Weight)		
	β	SE	*p*	β	SE	*p*	β	SE	*p*	β	SE	*p*
Model I												
rs855791 (G/A) **	−4.495	2.548	0.080	0.378	0.428	0.379	−0.232	0.146	0.133	−0.213	0.543	0.064
Iron Intake (mg)	1.476	0.551	0.009	−0.109	0.093	0.241	−0.018	0.031	0.565	0.059	0.104	0.518
Model II												
rs4820268 (A/G) †	−5.003	2.453	0.044	0.567	0.412	0.171	−0.137	0.142	0.335	−0.067	0.104	0.458
Iron Intake (mg)	1.518	0.549	0.007	−0.114	0.092	0.221	−0.017	0.032	0.598	−0.192	−1.111	0.745

Multiple Linear Regression method = enter (N = 121), All factors were adjusted by age and sex of children, Model 1: SF; R^2 = 0.064, sTfR; R^2 = 0.002, Hb; R^2 = 0.009, Model II: SF; R^2 = 0.072, sTfR; R^2 = 0.011, Hb; R^2 = −0.023, * Serum ferritin (SF) was adjusted by CRP >5 mg/L and AGP >1 mg/L, ** rs855791 GG = 0/AG = 1/AA = 2, † rs4820268 AA = 0/AG = 1/GG = 2.

4. Discussion

Our study reported the association of the *TMPRSS6* genes to the hemoglobin and iron status indicators in young children in the area where IDA is prevalent. To our knowledge, this is the first study that reported such an association in young children (<24 months. We also included simultaneously both genes and iron intake in our analysis, which had not been reported in previous studies. The important finding of this study was that polymorphism of the *TMPRSS6* gene SNP rs855791 and rs4820268 and iron intake both had an association with the iron status, particularly in the SF concentration.

The study population came from a specific tribe in Lombok Island, Indonesia (*Sasaknese)*, which is categorized as a close population. We found the HWE calculation, p + q = 1 which means that both allele (A) of rs85571 and (G) of rs4820268 fulfill the equilibrium condition [9,10].

Our study showed that the minor allele of TMPRSS6 rs4820268 (G) and rs8559791 (A) were associated with a lower SF but not with sTfR, Hb and body iron store. Serum ferritin reflects the iron status indicator, which is sensitive to the early stage of iron deficiency, i.e., iron depletion stage, whereas sTfR reflects a later stage of iron deficiency [11,12]. In the development of ID, sTfR may still be in the normal concentration during the iron depletion stage (during which SF already decreased) and iron-deficient erythropoiesis (during which serum iron and transferrin saturation start to decrease) [13].

This study was done within a specific ethnicity in Indonesia (*Sasaknese*) and did not represent the multi-ethnicity population of Indonesia. We consider *Sasaknese* as an example of a highly prevalent IDA population; and while this is a close population, the genotype data is typical of Asians. The Chinese Han population study by Gan et al. and another study by Chamber et al., *l* which include SNP rs855791 and rs4820268 of the *TMPRSS6* genes showed that minor allele frequency (MAF) of both SNPs in Asia was comparable with the finding in our study population (>55%) [5,14]. There is a considerable discrepancy in the minor allele frequency distribution of both SNPs among the Asian, African and European population [5,14,15]. The minor allele frequency of SNP rs855791 and rs4820268 tend to be lower in African than European and Asian populations. This difference may be caused by certain environmental conditions acting through selective pressure to alter the frequency of the genetic variants among population [16]. Our finding on the variant allele frequency showed that the frequency of variant allele in both SNPs among *Sasaknese* was higher than the study in the Chinese Han and Indian Asian population in which variant allele "A" of rs855791 and variant allele "G" of rs4820268 was within range 50–56% [5,14]. This condition could have occurred because *Sasaknese* had a tradition to maintain their lineage through marriage with the same *Sasaknese.* Thus, the probability of variant allele to appear in the next generation was higher than random mating.

Our finding showed that iron intake had a more significant association as compared to polymorphism of rs855791 and rs4820268 with SF concentration. These results showed that while both gene (nature) and intake (nurture) had an association with the iron status, nurture had a more significant role to the iron status in human. Tanaka et al. found that 20–30% of the variability in iron concentration is attributable to reduced activity of the *Matriptase-2* protein; the latter is determined by the environment including the dietary iron intake [3]. This finding is similar with the study in Tanzanian children, which found that the daily iron intake had a positive correlation to the SF level ($r = 0.233$, $p < 0.05$) [6]. A similar finding was found on a study among Indian children 12–23 months that according to the multiple linear regression analysis, ferritin was positively associated with iron intake (crude$\beta = 0.25$ [0.02–0.12]; $p < 0.01$) [17]. Nonetheless, in our study, the variant allele of rs855791 (A) and rs4820268 (G) were associated with a lower SF level even after controlling for iron intake.

Previous mutation on *TMPRSS6* has been shown to result in the *Matriptase-2* protein lacking activity, which is essential for adequate iron uptake to prevent iron deficiency and to suppress hepcidin expression. Hepcidin is a key iron regulator, which governs systemic iron homeostasis by binding it to ferroportin on the surface of macrophages, enterocyte, and hepatocyte inducing the degradation of ferroportin thereby preventing the efflux of iron in the blood [3,14,15]. In the present study, *TMPRSS6* showed that per copy of the minor allele (G) of rs4820268, the SF concentration reduced by 5.00 μg/L ($p = 0.044$). Meanwhile per copy of the minor allele (A) of rs85579, the SF concentration reduced by 4.50 μg/L ($p = 0.08$). A reduction of 1 μg/L in SF is considered significant as it corresponds to 8–10 mg body iron, while the total body iron store in humans is around 1000–1200 mg [18]. This means that those who have a minor homozygote allele genotype (AA) of rs855791 and (GG) of rs4820268 are naturally depleted by approximately 10% of the total body iron store as compared to the wild-type homozygote. This finding may suggest for the sub-population specific strategies to address iron deficiency, i.e., whether populations with a high prevalence of risk alleles may require higher nutrient recommendations for iron to maintain normal erythropoiesis. This is especially prudent where the high prevalence of risk alleles is combined with the high inflammation burden.

The limitation of our study was that we did not assess hepcidin. We hypothesized that hepcidin is lower in homozygote wild-type as found in Chinese women study where homozygote wild-type (GG) of rs855791 had a lower hepcidin concentration than homozygote minor allele genotype (AA) in the general population and therefore the reduction of the SF concentration by minor allele of rs855791 and rs4820268 was due to over expression of hepcidin [14]. On the other hand, another study by Guo et al. examined the correlation between the hepcidin concentration and iron status indicators and reported that there was a significant positive correlation between serum hepcidin and ferritin concentration among the Chinese Han Population [19]. Another limitation of our study is a small sample size in terms of the interaction between gene and iron biomarkers (statistical power ~70% based on correlation coefficient of 0.38 in SF with significance level of 0.05), therefore we acknowledge the need for a study with a bigger sample size to determine the exact nature of the interaction. Nevertheless, we managed to analyze simultaneously the influence of both gene and intake, which has not been reported before.

5. Conclusions

In summary, our study showed that amongst these under-two-year-old children, where anemia and iron deficiency are prevalent, variant allele (A) and (G) at rs855791 and rs4820268 of the *TMPRSS6* gene respectively were associated with a lower SF concentration; but the association was still weaker than that of iron intake. A further study is needed to investigate whether a higher nutrient recommendation for iron based on genetic characteristics, particularly for children who have mutation in *TMPRSS6* will benefit their iron status.

Author Contributions: U.F., M.K.H. and C.A. designed the research; U.F., M.K.H., C.A. and D.S. conducted the research; D.S. analyzed the data; D.S., U.F. and A. interpreted the statistical analysis; D.S., U.F., A. and M.K.H. wrote the manuscript. All authors read and approved the final manuscript.

Funding: This study was funded by the Ministry of Education Government of Indonesia (SEAMEO RECFON DIPA 2015), Scientific Programme Indonesia Netherlands Academy of Arts and Sciences (SPIN-KNAW) (Award Letter 09-PD-12), *Beasiswa Pendidikan Indonesia Lembaga Pengelolaan Dana Pendidikan (BPI-LPDP)* (no. 20160822028741).

Acknowledgments: The authors are grateful to Dwi Anita Suryandari, Luh Ade Wiradnyani for their suggestion for the study; Judhiastuty Februhartanty for language editing, Putu Amrytha Sanjiwani and Ria Amelia for their assistance in the laboratory. We would particularly like to thank all the participants, voluntary health workers and local leaders in Suralaga sub-district, East Lombok District, West Nusa Tenggara Indonesia for their cooperation.

Conflicts of Interest: The authors declare no conflict of interest.

References

1. WHO Global Database on Anaemia. *Worldwide Prevalence of Anaemia 1993–2005*; World Health Organization: Geneva, Switzerland, 2008.
2. McLean, E.; Cogswell, M.; Egli, I.; Wojdyla, D.; de Benoist, B. Worldwide prevalence of anaemia, WHO vitamin and mineral nutrition information system, 1993–2005. *Public Health Nutr.* **2009**, *12*, 444–454. [CrossRef]
3. Tanaka, T.; Roy, C.N.; Yao, W.; Matteini, A.; Semba, R.D.; Arking, D.; Walston, J.D.; Fried, L.P.; Singleton, A.; Guralnik, J.; et al. A genome-wide association analysis of serum iron concentrations. *Blood J.* **2010**, *115*, 94–96. [CrossRef]
4. Gichohi-Wainaina, W.N.; Towers, G.W.; Swinkels, D.W.; Zimmermann, M.B.; Feskens, E.J.; Melse-Boonstra, A. Inter-ethnic differences in genetic variants within the transmembrane protease, serine 6 (*TMPRSS6*) gene associated with iron status indicators: A systematic review with meta-analyses. *Genes Nutr.* **2014**, *10*, 442. [CrossRef]
5. Chambers, J.C.; Zhang, W.; Li, Y.; Sehmi, J.; Wass, M.N.; Zabaneh, D.; Hoggart, C.; Bayele, H.; McCarthy, M.I.; Peltonen, L.; et al. Genome-wide association study identifies variants in *TMPRSS6* associated with hemoglobin levels. *Nat. Genet.* **2009**, *41*, 1170–1172. [CrossRef] [PubMed]
6. Tatala, S.; Ndossi, G.; Ash, D. Impact of dietary iron intake on anaemia in Tanzanian schoolchildren. *Sajcn* **2004**, *17*, 94–100. [CrossRef]
7. Thurnham, D.I.; Mccabe, G.P. Influence of infection and inflammation on biomarkers of nutritional status with an emphasis on vitamin A and iron. In *World Health Organization, Proceedings of the Priorities in the Assessment of Vitamin A and Iron Status in Populations, Panama City, Panama, 15–17 September 2010*; World Health Organization: Geneva, Switzerland, 2012.
8. Cook, J.D.; Skikne, B.; Baynes, R. The use of transferrin receptor for the assessment of iron status. In *Iron Nutrition in Health and Disease*; Hallberg, L.A., Asp, N.-G., Eds.; John Libbey& Co.: London, UK, 1996; pp. 91–99.
9. James, W.H. The validity of Weinberg's differential rule. *Twin Res. Hum. Genet.* **2007**, *10*, 771–772. [CrossRef]
10. Mayo, O. A century of Hardy-weinberg equilibrium. *Twin Res. Hum. Genet.* **2008**, *11*, 249–256. [CrossRef]
11. Gibson, S.R. *Principles of Nutritional Assesment*, 2nd ed.; Oxford University Press: New York, NY, USA, 2005; 446p.
12. Thurnham, D.I.; Mburu, A.S.W.; Mwaniki, D.L.; Wagt, A.D. Micronutrients in childhood and the influence of subclinical inflammation. *Proc. Nutr. Soc.* **2005**, *64*, 502–509. [CrossRef]
13. Åkesson, A.; Bjellerup, P.; Berglund, M.; Bremme, K.; Vahter, M. Serum transferrin receptor: A specific marker of iron deficiency in pregnancy. *Am. J. Clin. Nutr.* **1998**, *68*, 1241–1246. [CrossRef] [PubMed]
14. Gan, W.; Guan, Y.; Wu, Q.; An, P.; Zhu, J.; Lu, L.; Jing, L.; Yu, Y.; Ruan, S.; Xie, D.; et al. Association of *TMPRSS6* polymorphisms with ferritin, hemoglobin, and type 2 diabetes risk in a Chinese Han population. *Am. J. Clin. Nutr.* **2012**, *95*, 626–632. [CrossRef] [PubMed]
15. Gichohi-Wainaina, W.N.; Melse-Boonstra, A.; Swinkels, D.W.; Zimmermann, M.B.; Feskens, E.J.; Towers, G.W. Common variants and haplotypes in the TF, TNF-alpha, and *TMPRSS6* genes are associated with iron status in a female black South. *J. Nutr.* **2015**, *145*, 945–953. [CrossRef] [PubMed]
16. Fumagalli, M.; Sironi, M.; Pozzoli, U.; Ferrer-Admettla, A.; Pattini, L.; Nielsen, R. Signatures of environmental genetic adaptation pinpoint pathogens as the main selective pressure through human evolution. *PLoS Genet.* **2011**, *7*, e1002355. [CrossRef]
17. Tatala, S.; Svanberg, U.; Mduma, B. Low dietary iron availability a major cause of anaemia. A nutrition survey in Lindi District of Tanzania. *Am. J. Clin. Nutr.* **1998**, *68*, 171–178. [CrossRef] [PubMed]

18. Zimmermann, M.B.; Hurrell, R.F. Nutritional iron deficiency. *Lancet* **2007**, *370*, 511–520. [CrossRef]
19. Guo, X.; Zhou, D.; An, P.; Wu, Q.; Wang, H.; Wu, A.; Mu, M.; Zhang, D.; Zhang, Z.; Wang, H.; et al. Associations between serum hepcidin, ferritin and Hb concentrations and type 2 diabetes risks in a Han Chinese population. *Br. J. Nutr.* **2017**, *110*, 2180–2185. [CrossRef] [PubMed]

nutrients

MDPI

Article

Estimation of the Burden of Iron Deficiency Anemia in France from Iron Intake: Methodological Approach

Juliana De Oliveira Mota [1], Patrick Tounian [2], Sandrine Guillou [1], Fabrice Pierre [3] and Jeanne-Marie Membré [1,*]

1 Secalim, INRA, Oniris, Université Bretagne Loire, 44307 Nantes, France
2 Service de nutrition et gastroentérologie pédiatriques, hôpital Trousseau, Sorbonne Université, AP–HP, 75012 Paris, France
3 INRA, ToxAlim (Research Centre in Food Toxicology), Université de Toulouse, INRA, ENVT, INP-Purpan, UPS, 31027 Toulouse, France
* Correspondence: jeanne-marie.membre@oniris-nantes.fr; Tel.: +33-240-684-058

Received: 24 July 2019; Accepted: 27 August 2019; Published: 1 September 2019

Abstract: Dietary iron deficiency (ID) is the first nutritional deficiency in the world, in terms of disability adjusted life years (DALY). This nutritional deficiency may lead to anemia, especially among children, adolescents, and adult women. The aim of this study was to build an original probabilistic model to quantitatively assess the ID, the iron deficiency anemia (IDA) and the subsequent health burden in France expressed in DALY, per age class and gender. The model considered the distribution of absorbed iron intake, the iron requirement distribution established by the European Food Safety Authority and the iron status in France. Uncertainty due to lack of data and variability due to biological diversity were taken into account and separated using a second-order Monte Carlo procedure. A total of 1290 (95% CI = 1230–1350) IDA cases corresponding to 16 (95% CI = 11–20) DALY were estimated per 100,000 individuals per year. The major contributors to IDA burden were menstruating females aged from 25 to 44 years old. Then, a consumption scenario was built with ground beef as intake, an increase in red meat consumption to 100 g/d would not eliminate entirely the IDA burden. The quantitative methodology applied here for France could be reused for other populations.

Keywords: iron deficiency; anemia; risk assessment; probabilistic model; second-order Monte Carlo simulation; DALY

1. Introduction

The World Health Organization (WHO) estimated that around two billion people were anemic in the world, which makes this disease one of the most common nutrition disorders [1]. Anemia was defined as a hemoglobin (Hb) concentration below the thresholds given by WHO, United Nations Children's Fund (UNICEF), United Nations University (UNU) [2]. For children aged from 6 to 59 months of age and pregnant women, anemia is diagnosed when Hb is under 11 g/dL, under 11.5 g/dL for children aged 5–11, and under 12 g/dL for children aged 12–14 and non-pregnant females. Finally, for males aged 15 years and above, anemia is diagnosed when the level of Hb is under 13 g/dL. The main cause of anemia identified was iron deficiency (ID), which is defined by WHO as "a state in which there is insufficient iron to maintain the normal physiological function of tissues such as the blood, brain, and muscles" [1]. ID is characterized by no mobilized iron stores and signs of a compromised supply of iron [3]. ID was estimated to represent over than 60% of the anemia causes [4,5], especially among the younger population [4,6]. This nutritional deficiency—commonly defined as serum ferritin level on blood $\leq$30 μg/L [7,8]—has several consequences for human health, such as a delay in children's mental and physical development, a decrease in work productivity, intelligence or cognitive capacity [1,9]. In some countries, ID and iron deficiency anemia (IDA) may also contribute to child, maternal, and perinatal mortality [1,9]. Among the overall population, the most concerned by this nutritional disease are children, adolescents, and women (pregnant and non-pregnant) in reproductive age [4,5,10].

To reduce ID, the consumption of bioavailable iron-rich food—such as red meat—would be a logically feasible solution, since this food is the richest source of high absorbable iron (heme-iron) [11]. A study revealed that an increase in red meat consumption twice a week compared to seldom intake would decrease 24% of child ID in Israel [12]. Another study estimated an increase of 0.6% in serum ferritin in blood levels per 1g of red meat consumed [13].

Since 1990, the Global Burden of Diseases (GBD) study, supported by the World Health Organization, has estimated the disability adjusted life years (DALY) associated with health effects of major diseases, injury, and risk factors. This metric takes into account mortality and morbidity and is expressed as the sum of the number of years of life lost (YLL) from premature death and the number of years lived with disability (YLD) [14]. In addition to quantifying the impact of disease, the burden of diseases helps in risk-mitigating strategy: Several consumption patterns can be evaluated and compared by scenario analysis. In particular, in the study of Kassebaum on behalf of GBD 2013 Anemia Collaborators evaluated the global burden of anemia [4]. This work estimated that, in 2013, anemia was responsible for 61.5 (41.0–88.7) millions of YLD in the world, from which 300,654 (200,463–434,060) YLD in France. However, this burden was estimated from hemoglobin levels, which makes scenarios of iron consumption impossible.

Regarding the separation of uncertainty—due to the lack of data and knowledge and variability associated with heterogeneity within populations—[15,16] of the inputs, it is important to identify which of them are driving the output of the risk model and to identify the needed data to increase the precision and the confidence of the estimated output [17].

The objective of this study was to estimate the burden of disease in France due to ID, taking iron consumption into account. Based upon dietary surveys, iron recommendations and blood iron status of the French population, a probabilistic model was set up for both genders and for specific age classes to quantify the risk of ID and the consequent burden of disease due to IDA. Then, to reduce the burden, consumption scenario was built with ground beef intake as red meat is the richest source of high absorbable iron in developed countries.

2. Methods

2.1. Model Framework for Iron Deficiency Anemia Assessment

The model was developed for males and females aged from three years old for which the age classes were 3–6, 7–11, 12–17, 18–24, 25–44, 45–64, and 65–74. For female population, the menstruating and menopausal populations were identified as follows: Between 15 and 24 years old, all women are menstruating, between 25 and 64 there are both menstruating and menopausal women and from 65 onwards there are only menopausal women. In addition, for adolescent females, two age classes were identified: 12–14 which were considered non menstruating, and 15–17 which were considered as menstruating.

The flowchart of the risk assessment model is presented in Figure 1. From iron consumption in France, the mean absorption of iron, the iron recommendations, and hemoglobin level in the French population, the number of cases and then the burden of disease per year, gender, and age class were estimated.

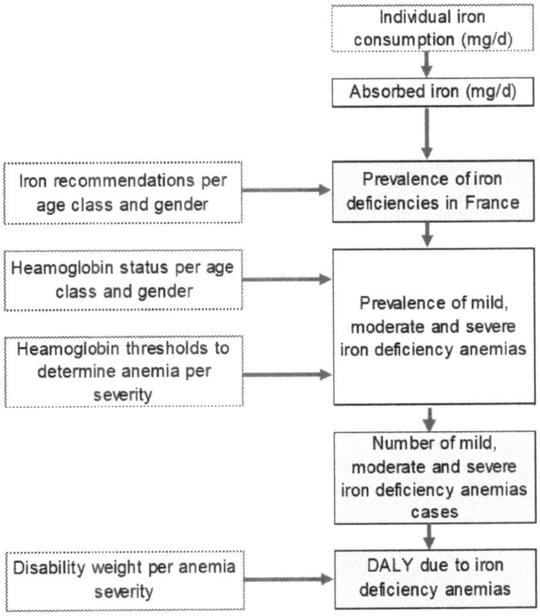

Figure 1. Flowchart of the assessment model of iron deficiency anemia disease per year in France per age class and gender. White rectangles with dashed line correspond to the "Inputs", full line to "Intermediate calculation". Light grey rectangles correspond to the "Final output". Absorbed iron corresponds to the mean absorbed values provided by European Food Safety Authority (EFSA) considering both heme and non-heme iron.

When building the model, several assumptions were made and validated by a medical expert. Three hypotheses were considered:

- ID appears when iron requirements are not covered by dietary iron.
- The ratio between the proportion of ID and IDA remains constant over the consumption scenarios.
- The consumption of bioavailable dietary iron, such as that found in red meat, reduces IDA.

2.2. *Iron Intake Consumption in France*

Iron intake in metropolitan France was evaluated between 2005 and 2007 for males and females aged from 3 to 79, by the dietary survey INCA 2 (approved by the French National Commission for Computed Data and Individual Freedom, "Commission Nationale Informatique et Libertés"; CNIL, under the registration code 797859—v0, on 20 March 2002). From the original data available at https://www.data.gouv.fr/ [18], we selected all-male population and non-pregnant women whose information about pregnancy and menopausal status was available. For unknown status of adult women, this latter was deduced or hypothesized. When the answer about menopausal status was unknown, the pregnancy status was verified. If the latter was negative, the average age of menopausal occurrence was taken into account. Therefore, women aged over 50 years old were assumed menopausal. Women aged less or equal to 50 years old, were assumed pre-menopausal [19]. On the other hand, when pregnancy status was unknown, the menopausal status was verified. If the answer was negative, women were considered as pregnant. Pregnant women were excluded from the study because the number of cases was low for each age class concerned (under 30). Under 18 years old, women were considered not pregnant. From the iron consumed, only a certain amount was absorbed. As for

European Food Safety Authority (EFSA) estimations, the absorbed iron from iron intake was 10% for children under 11 years old, 16% for adolescents and adult males, and 18% for adult females [20].

To estimate the probability density of iron intake, the function fitdist of the package fitdistrplus was used with the R software version 3.4.0. Based upon the Akaike information criterion (AIC), the lognormal distribution provided the best fit, among Weibull, normal, lognormal and gamma distributions.

2.3. Prevalence of Iron Deficiency in France

To calculate ID prevalence in France, iron intake from dietary survey INCA 2 (see Table 1 at "Iron consumption (absorbed)"), was compared with a normal distribution of iron needs, as suggested by the Nordic Council of Ministers in 2014 [21] based on values given by the European Food Safety Authority (EFSA) (2015) [20] (see Table 1 at "Iron needs"). The mean parameter of the normal distribution was the absorbed average requirement value, the standard deviation was the value directly provided by EFSA report, except for males under 17 and females under 14 for which the standard deviation was deduced from the confidence interval upper limit bound of the EFSA report. In the present study, adolescent girls were divided into two groups: Girls from 12 to 14 were considered having not a menstruating status—mean menarcheal age of 13 years old in France [22]—and girls aged from 15 to 17 were assumed to have the iron needs as adult women since all the individuals in this age class were assumed to be menstruating.

The prevalence of people consuming less than the normal probability distribution of the needs was then estimated, using plnorm function of R software. This population was considered as iron-deficient. By multiplying the prevalence of ID in France by the number of individuals for each age class and gender (see Table 1 at "French population data"), the number of ID cases in France per year was estimated.

2.4. Prevalence of Iron Deficiency Anemia in France

Anemia results from levels of hemoglobin <12–15 g/dL and 51% (21%–85%) of anemia is due to ID [10,12,23]. To estimate IDA prevalence from iron deficiency, we based our calculations on the reported iron deficiencies, iron deficiencies anemias, and anemias from three studies [24–26]. From ID, IDA were calculated as follows:

$$Prev.IDA_{a,g} = Prev.ID_{a,g} \times Prop.IDA_{a,g} \quad (1)$$

where a is the age class, g the gender, *Prev.IDA* the prevalence of IDA, *Prev.ID* the prevalence of ID and *Prop.IDA* the proportion of IDA due to ID in France. The data are given in Table 1.

To determine the proportion of anemias per severity (mild, moderate, severe), the level of hemoglobin of the French population was considered. Stoltzfus et al., 2004 provided hemoglobin levels in blood of children in EUR-A WHO region. The standard deviation of hemoglobin taken into account in this study was 1.0 g/dL for countries with anemia prevalence under 15%, as it is in France [9,23] (see Appendix A.1 Table A1). In 2011, a study evaluated the iron status in European adolescents and found similar hemoglobin levels [27]. The following age class and hemoglobin levels were estimated from linear functions. Adult male and female hemoglobin levels were provided by Santé Publique France data from ENNS (Étude Nationale Nutrition Santé) study [26]. The methodology to determine hemoglobin levels of the French population is presented in Appendix A.1 and the levels used in the present study are given in Table 1 and in Appendix A.1 Table A2. These levels of hemoglobin followed a normal distribution from which the proportion of anemia in France (with and without ID) was estimated.

Table 1. Sources of information and implementation of the inputs either as deterministic values or as probability distributions.

Characteristic	Input	Gender [1]	Model Implementation Per Age Class [2] 3–6	7–11	12–14	15–17	18–24	25–44	45–64	65–74	Unit	Type [3]	From
French population data	*Pop*	Male	1,571,427	1,925,359	2,352,805		2,788,141	8,279,094	7,663,979	2,269,631			
		Female Pre-M	1,498,259	1,829,236	1,084,687	1,159,862	2,649,398	7,910,584	2,694,011	-	number	D	[28]
		Female Post-M	-	-	-	-	-	157,425	5,313,189	2,712,349			
Iron consumption (absorbed)		Male	$LogN(-0.21\ [95\%\ CI = -0.26 -(-0.17)],\ 0.31\ [95\%\ CI = 0.27 -0.36])$	$LogN(0.06\ [95\%\ CI = 0.01 -0.09],\ 0.28\ [95\%\ CI = 0.26 -0.31])$	$LogN(0.62\ [95\%\ CI = 0.59 -0.66],\ 0.34\ [95\%\ CI = 0.31 -0.36])$		$LogN(0.52\ [95\%\ CI = 0.46 -0.58],\ 0.37\ [95\%\ CI = 0.33 -0.42])$	$LogN(0.75\ [95\%\ CI = 0.71 -0.78],\ 0.31\ [95\%\ CI = 0.29 -0.34])$	$LogN(0.74\ [95\%\ CI = 0.71 -0.78],\ 0.32\ [95\%\ CI = 0.30 -0.34])$	$LogN(0.79\ [95\%\ CI = 0.75 -0.84],\ 0.34\ [95\%\ CI = 0.31 -0.37])$			
		Female Pre-M	$LogN(-0.29\ [95\%\ CI = -0.34 -(-0.25)],\ 0.25\ [95\%\ CI = 0.21 -0.28])$	$LogN(-0.07\ [95\%\ CI = -0.10 -(-0.03)],\ 0.29\ [95\%\ CI = 0.26 -0.31])$	$LogN(0.43\ [95\%\ CI = 0.38 -0.48],\ 0.34\ [95\%\ CI = 0.30 -0.37])$	$LogN(0.40\ [95\%\ CI = 0.36 -0.45],\ 0.36\ [95\%\ CI = 0.33 -0.40])$	$LogN(0.52\ [95\%\ CI = 0.46 -0.58],\ 0.37\ [95\%\ CI = 0.33 -0.42])$	$LogN(0.59\ [95\%\ CI = 0.56 -0.62],\ 0.35\ [95\%\ CI = 0.33 -0.37])$	$LogN(0.67\ [95\%\ CI = 0.62 -0.73],\ 0.39\ [95\%\ CI = 0.35 -0.43])$	-	mg/day	U and V	[18]
		Female Post-M	-	-	-	-	-	$LogN(0.19\ [95\%\ CI = -0.09 -0.47],\ 0.43\ [95\%\ CI = 0.25 -0.63])$	$LogN(0.64\ [95\%\ CI = 0.60 -0.68],\ 0.38\ [95\%\ CI = 0.35 -0.41])$	$LogN(0.58\ [95\%\ CI = 0.53 -0.64],\ 0.34\ [95\%\ CI = 0.30 -0.38])$			
Iron needs	*IR*	Male	$N(0.5, 0.1)$	$N(0.8, 0.16)$	$N(1.27, 0.25)$		$N(0.97, 0.38)$	$N(0.97, 0.38)$	$N(0.97, 0.38)$	$N(0.97, 0.38)$			
		Female Pre-M	$N(0.5, 0.1)$	$N(0.8, 0.16)$	$N(1.13, 0.23)$	$N(1.41, 0.76)$	$N(1.41, 0.76)$	$N(1.41, 0.76)$	$N(1.41, 0.76)$		mg/day	V	[20]
		Female Post-M	-	-	-	-		$N(0.97, 0.38)$	$N(0.97, 0.38)$	$N(0.97, 0.38)$			
Proportion of iron deficiencies anemia when iron deficiency	*Prop.IDA*	Male	0.147	0	0		0	0.26	0	0.42			
		Female Pre-M	0.147	0	0	0.5	0.05	0.09	0.10		number	D	[25,26]
		Female Post-M	-	-	-	-	-	0	0	0.03			
Hemoglobin status	*Hb*	Male	$N(12.8, 1)$	$N(13.3, 1)$	$N(13.9, 1)$		$N(15.5, 0.8)$	$N(15.3, 1)$	$N(15.3, 1.1)$	$N(14.9, 1.9)$			
		Female Pre-M	$N(12.7, 1)$	$N(13.4, 1)$	$N(13.4, 1)$	$N(13.4, 1)$	$N(13.5, 1.5)$	$N(13.5, 1.4)$	$N(13.7, 1.1)$		g/dL	V	[23,26]
		Female Post-M	-	-	-	-	-	$N(13.8, 0.6)$	$N(13.8, 0.9)$	$N(14, 1.2)$			
Hemoglobin threshold for anemia severity: mild	*Hb.t*	Male	10.9	11.4	11.9	11.9	12.9	12.9	12.10	12.9	g/dL	D	[4]
		Female	10.9	11.4	11.9								
moderate		Male	9.9	10.9	10.9				10.10	10.9			
		Female	9.9	10.9									
severe		Male	7	8					9	8			
		Female	7	8									
Disability weight: mild	*w*	Both	$N(0.005, 0.002)$										
moderate		Both	$N(0.058, 0.012)$								number	U	[29]
severe		Both	$N(0.164, 0.030)$										

[1] Female Pre-M, Premenopausal females; Female Post-M, Postmenopausal females. Menstruating females considered between 15 years and 64 years at most; [2] Following R parametrization; [3] D, deterministic; V, Variability; U, uncertainty.

According to "The burden of diseases of anemia", three levels of anemia were considered: Severe, moderate, and mild, as used by Kassebaum on the behalf of GBD 2013 Anemia Collaborators evaluation (2016) based on which health state descriptions are presented in Table 2.

Table 2. Iron deficiency anemia severity levels and health state descriptions from the Global burden of disease 2013 study [29].

Health State	Health State Description
Mild IDA	Feels slightly tired and weak at times, but this does not interfere with normal daily activities.
Moderate IDA	Feels moderate fatigue, weakness, and shortness of breath after exercise, making daily activities more difficult.
Severe IDA	Feels very weak, tired and short of breath, and has problems with activities that require physical effort or deep concentration.

The prevalence of anemia per severity, $Prev.IDA_{a,g,s}$, was then calculated as follows:

$$Prev.IDA_{a,g,s} = Prev.IDA_{a,g} * Alloc.Ane_{a,g,s} \tag{2}$$

The allocation of anemia severity, among the anemia population, $Alloc.Ane_{a,g,s}$, is deduced from the hemoglobin normal distribution per age class and gender (*Hb*) in the French population and hemoglobin thresholds (*Hb.t*) defined by WHO [4]. The methodology is provided in the Appendix A Method A2. We assumed that the proportion of anemia per severity (with and without ID) in the French population was the same as for anemia only due to ID.

2.5. Burden of Diseases from Iron Deficiency Anemia

The burden of diseases of the outcome attributable to IDA was expressed in the number of cases and disability adjusted life years (DALY). This metric estimates an equivalent number of years in good health lost due to the outcome [30].

DALY estimation takes into account mortality by considering the YLL for the age class and gender concerned. In addition, morbidity is also included and expressed in YLD. YLD is obtained by multiplying the decrease in the quality of life, expressed by a disability weight factor due to the sequelae or the stage of the outcome, by the duration of the specific disability. The disability weight factor is estimated between 0 (perfect health state) and 1 (death), available in the GBD estimations [4,29].

DALY estimation was calculated for the French population per gender and age class and expressed per 100,000 people per year, for both outcomes.

Number of IDA per age class, gender and severity ($NB.IDA_{a,g,s}$) was calculated, considering the number of individuals in the concerned population ($Pop_{a,g}$):

$$NB.IDA_{a,g,s} = Pop_{a,g} \times Prev.IDA_{a,g,s} \tag{3}$$

Then, the number of DALY due to IDA per age class and gender ($DALY_{a,g}$) was estimated as follows:

$$DALY_{a,g} = \sum_{s} NB.IDA_{a,g,s} \times w_{a,g,s} \tag{4}$$

where w was the disability weight factor.

2.6. Second-Order Monte Carlo Simulation

To estimate the ID prevalence, two distributions were used: Iron intake and iron needs. The iron intake distribution came from a fitting procedure with bootstrapping to separate uncertainty and variability. The iron needs' normal distribution was built using EFSA report values, it included only variability. The ID prevalence was then estimated by Monte Carlo simulation in two dimensions

(MC2D). One thousand iterations were run to capture the uncertainty and 10,000 iterations for the variability. Note that once the ID prevalence was estimated with its uncertainty, the outputs of the model conveyed only uncertainty. To verify the stability of the outputs, three simulations were carried out for each age class and gender, a variation less than 1% for the DALY output was obtained.

2.7. Consumption Scenarios

To decrease ID and IDA for the most concerned population—all populations with IDA—consumption scenarios were performed by setting the consumption of a food product rich in iron, at different quantities. Red meat was chosen as food model because it is the richest source of heme-iron in developed countries [11], which is the most absorbable type of iron by human beings, with 20% to 30% of iron intake absorbed [10]. The consumption scenario was made for consumption of cooked ground beef with 15% of fat, which is the type of beef meat most consumed by the French population (51 g per week) [18], most acceptable by youngest populations and easiest to eat by elderly populations. From initial iron intake by French individuals involved in INCA 2 study, and taken into account in this study, the iron from unprocessed red meat (muscle of beef, pork, veal, horse, lamb) consumption was estimated from CIQUAL tables available at https://ciqual.anses.fr/ [31] (see Appendix A.2 Table A3). The amount of iron from red meat consumption was subtracted from the total iron intake for the consumption scenario 0g/d of red meat. The consumption of 25 g/d, 50 g/d, 75 g/d, 100 g/d of cooked ground beef 15% fat was applied, with an amount of 2.6 mg of iron per 100 g of meat and an average absorption of 25% [10]. The consumption scenario took the amount of iron from other dietary sources (without red meat) and the addition of the amount of ground beef into account.

3. Results

3.1. Prevalence and Number of Iron Deficiencies in France

The distributions of ID with current consumption of iron in France are shown in Table 3. The mean number of ID cases is represented with its 95% confidence interval.

Table 3. Prevalence and number of iron deficiencies per 100,000 French individuals, per age class and gender. Mean value and its 95% confidence interval.

<table>
<tr><th rowspan="3">Age Class</th><th colspan="2" rowspan="2">Male</th><th colspan="4">Female</th></tr>
<tr><th colspan="2">Premenopausal</th><th colspan="2">Postmenopausal</th></tr>
<tr><th>Prevalence</th><th>Number of Cases</th><th>Prevalence</th><th>Number of Cases</th><th>Prevalence</th><th>Number of Cases</th></tr>
<tr><td>3–6</td><td>9% (6%–12%)</td><td>220 (140–310)</td><td>9% (7%–13%)</td><td>230 (160–300)</td><td>-</td><td>-</td></tr>
<tr><td>7–11</td><td>20% (16%–24%)</td><td>620 (510–730)</td><td>31% (27%–36%)</td><td>930 (810–1100)</td><td>-</td><td>-</td></tr>
<tr><td>12–14</td><td rowspan="2">15% (13%–18%)</td><td rowspan="2">580 (490–680)</td><td>21% (17%–25%)</td><td>360 (290–440)</td><td>-</td><td>-</td></tr>
<tr><td>15–17</td><td>43.7% (41%–47%)</td><td>820 (760–880)</td><td>-</td><td>-</td></tr>
<tr><td>18–24</td><td>9% (6%–12%)</td><td>400 (280–530)</td><td>37% (33%–41%)</td><td>1600 (1400–1700)</td><td>-</td><td>-</td></tr>
<tr><td>25–44</td><td>5% (4%–7%)</td><td>720 (580–880)</td><td>32% (30%–34%)</td><td>4100 (3900–4400)</td><td>34% (17%–50%)</td><td>90 (40–130)</td></tr>
<tr><td>45–64</td><td>4% (3%–4%)</td><td>450 (360–460)</td><td>27% (24%–31%)</td><td>1200 (1100–1300)</td><td>8% (7%–10%)</td><td>720 (580–860)</td></tr>
<tr><td>65–74</td><td>3.4% (2%–5%)</td><td>130 (80–190)</td><td>-</td><td>-</td><td>10% (7%–12%)</td><td>420 (320–530)</td></tr>
</table>

Among children, the most concerned with ID were children aged from 7 to 11, with 20% (95% CI = 16%–24%) and 31% (95% CI = 27%–36%) ID in this age class for males and females, respectively. Adolescent females were also more concerned with ID compared to adolescent males (Table 3).

The higher impact of ID among females was also noted in adults' age classes. Menstruating adult females aged from 18 to 24 were the most concerned with ID with 37% [95% CI = 33%–41%] of iron-deficient in this population group, corresponding to 1600 [95% CI = 1400–1700] IDs cases per 100,000 individuals. After this age class, the prevalence of ID decreases with increasing age.

3.2. Number of Iron Deficiency Anemias

In France, only children aged from three to six, older adolescent females, adult females and males aged from 25 to 44 years and from 65 to 74 suffered from IDA [25,26]. The distributions of IDA with current consumption of iron in France are shown in Table 4. The mean number of IDA cases is represented with its 95% confidence interval.

Table 4. Number of cases and disability adjusted life years (DALY) per 100,000 individuals of iron deficiencies anemias in France, per age class and gender. Mean value and its 95% confidence interval.

Age Class	Male		Female			
			Premenopausal		Postmenopausal	
	Number of Cases	DALY	Number of Cases	DALY	Number of Cases	DALY
3–6	32 (21–45)	0.3 (0.1–0.5)	33 (23–45)	0.3 (0.1–0.5)	-	-
7–11	0 (0–0)	0 (0–0)	0 (0–0)	0 (0–0)	-	-
12–14	0 (0–0)	0 (0–0)	0 (0–0)	0 (0–0)	-	-
15–17			404 (377–431)	4.0 (2.1–5.9)	-	-
18–24	0 (0–0)	0 (0–0)	77 (69–85)	1.6 (1.0–2.1)	-	-
25–44	183 (146–224)	0.9 (0.1–1.8)	374 (353–394)	6.5 (4.2–9.0)	0 (0–0)	0 (0–0)
45–64	0 (0–0)	0 (0–0)	122 (108–136)	1.2 (0.7–1.8)	0 (0–0)	0 (0–0)
65–74	54 (33–80)	0.6 (0.3–1.0)	-	-	11 (9–15)	0.1 (0.1–0.2)

A total of 1290 [95% CI = 1230–1350] cases of IDA per 100,000 individuals per year was estimated in the French population.

Children aged from three to six, had almost the same number of IDA cases for males and females with 32 (95% CI = 21–45) and 33 (95% CI = 23–45) cases per 100,000 individuals per year, respectively. Among adolescents, only females aged from 15 to 17 are concerned with IDA. The number of IDA cases for this population was ten times higher than for females aged from three to six.

Adult females were significantly more concerned with IDA than males with 584 (95% CI = 558–612) IDA cases versus 237 (95% CI = 194–282) IDA cases per 100,000 individuals, respectively. The major number of cases concerned menstruating women aged from 15 to 17 with 404 (95% CI = 377–431) IDA cases per 100,000 individuals per year. The cases of IDA for menopaused females only concerned the last age class.

3.3. DALY Attributable to Iron Deficiency Anemias

The number of DALY per 100,000 individuals per year associated with IDA in France is given in Table 4. The mean number of IDA DALY is represented with its 95% confidence interval.

The major contributors to the burden from IDA were menstruating females aged from 25 to 44 with 6.5 (95% CI = 4.2–9.0) DALY associated with IDA per 100,000 individuals per year. Young children aged from three to six, had a low burden with an equivalent number of DALYs (0.3 (95% CI = 0.1–0.5)

DALY per 100,000 per year). Older adolescent females, aged from 15 to 17, contributed with 4.0 (95% CI = 2.1–5.9) DALY per 100,000 individuals per year.

The estimated number of DALY associated with IDA was 16 (95% CI = 11–20) per 100,000 French individuals per year, from which more than 80% of the DALY were attributable to menstruating females.

3.4. Consumption Scenarios

Based on the developed model, to estimate the effect of red meat intake on IDA burden of diseases, a consumption scenario for young children (three to six), older adolescent females (15–17), males (25–44; ≥60), adults menstruating females (≥18) and menstruated females (65–74) was generated. The results for young children and older adolescent females are presented in Figure 2 and results for adult females are reported in Figure 3.

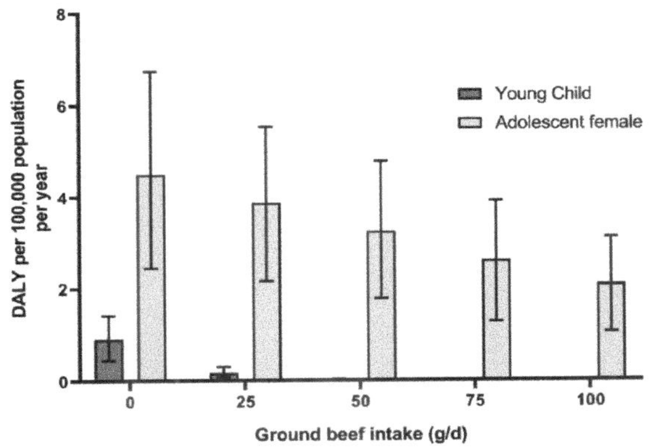

Figure 2. Estimated number of DALY from iron deficiency anemias in France for young children (three to six) and adolescent females (15–17) according to ground beef consumption scenarios. Results expressed per 100,000 individuals per year. Full lines represent the 95% uncertainty around the mean value.

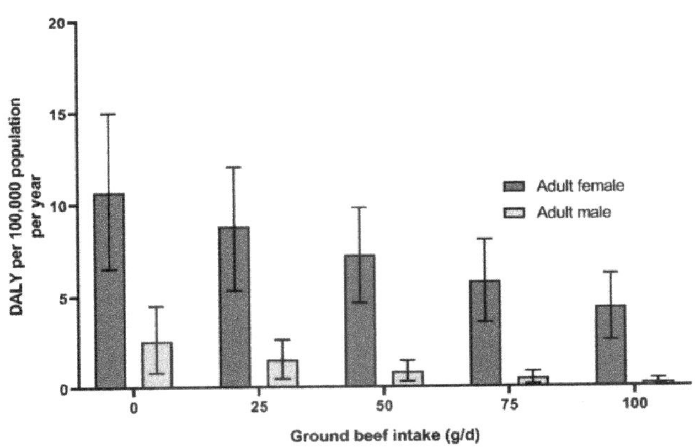

Figure 3. Estimated number of DALY from iron deficiency anemias in France for adult males and females according to ground beef consumption scenarios. Results expressed per 100,000 individuals per year. Full lines represent the 95% uncertainty around the mean value.

For young children, the consumption of 50 g/d of ground beef, added to the iron consumption from other sources of iron without red meat, in current diet, would be enough to reduce the prevalence of ID to 0.2% (95% CI = 0%–1.9%) per 100,000 individuals per year and to decrease the burden of IDA close to zero DALY. For older adolescent females, the consumption of 100 g/d of red meat—which may represent a portion of red meat per day—added to the other sources of iron in current diet, would not be enough to eliminate ID and IDA. However, it would reduce the prevalence of ID to 22% (95% CI = 0%–90%) per 100,000 individuals per year and halve IDA burden when comparing with current consumption (mean red meat consumption of 18.2 g/d).

The burden of IDA in male adults was already low, but the consumption of 50 g/d or 100 g/d of red meat, in addition to the other sources of iron in current diet (without red meat), would decrease the current number of DALY from 0.8 [95% CI = 0.3–1.4] and 0.2 [95% CI = 0.1–0.4] per 100,000 individuals per year, respectively.

The burden of IDA in adult females aged over 18, were the highest in the population, especially for menstruating females (Figure 3). The consumption of 50 g/d or 100 g/d of red meat, in addition to the other sources of iron in current diet (without red meat) would decrease the current number of DALY down to 7.2 (95% CI = 4.6–9.8) and 4.4 (95% CI = 2.5–6.2) per 100,000 individuals per year, respectively.

4. Discussion

The present study objective was to build an original probabilistic assessment model to estimate the burden of diseases associated with ID and subsequently IDA with the current iron consumption. A probabilistic risk assessment model was developed to estimate the prevalence of iron deficiencies, the number of cases and DALY in France from iron consumption.

Our results were in line with a previous study in France, from 2001, especially for young children [25]. According to the study, 13.6% of young children aged from two to six were iron-deficient, which is not far from our estimations where 9% (95% CI = 6%–12%) of young boys and 9% (95% CI = 7%–13%) of young girls were iron-deficient. However, there is a missing population in this comparison. Due to the lack of consumption data, our estimations only included children aged from three to six. This estimation was also in line with iron deficiencies prevalence in Europe estimated between 7% and 18% for toddlers and young children [32]. On the other hand, these results were higher than those found in France in a more recent study which found 3.2% (95% CI = 2.0%–5.1%) [33], although the threshold for ID definition in this study was low with 10 or 12 μg/L—corresponding to the total depletion of iron storage. Indeed, the definition of ID was not the same because, with the data available, it was not possible to quantify the total depletion of iron, but the iron under the average requirement. For adolescents, it was estimated 15% (95% CI = 13%–18%) ID prevalence for males aged from 12 to 17 years old and 21% (95% CI = 17%–25%) and 44% (95% CI = 41%–47%) ID prevalence for adolescent females aged from 12 to 14 and from 15 to 17, respectively. This was in line with the estimations in Europe (between 0% and 43%) but was higher than iron deficiencies estimated for French adolescents (0% for males aged from 14 to 18, 3.1% for females aged from 11 to 15 and 15.4% for females aged from 14 to 18) [24]. Nonetheless comparison was hard to establish because age classes were not the same as those in this study, the data were from 1988 and the definition of ID threshold was not the same [34]. For the adult population, the most concerned with ID were menstruating adult females. This is mainly because iron losses are higher than for males and menopausal females, then, iron needs are higher for menstruating females (1.41 mg/day) compared to other adults' needs (0.97 mg/day). In addition, iron consumption by adult menstruating females (mean 11 mg/d, sd 4 mg/d) was lower than by adult males (mean 14 mg/d, sd 5 mg/d). The values estimated in our study were similar to those measured in blood of French adults in ENNS study (for serum ferritin ≤30 μg/L) [35].

The estimated number of IDA cases was 1290 (95% CI = 1130–1350) IDA per 100,000 individuals per year (Table 4) with 584 (95% CI = 558–612) cases per 100,000 people for adult women per year. This corresponded to 797,200 (95% CI = 759,900–836,900) for the French population per year. The major contributors were menstruating females aged from 15 to 17. Indeed, twenty-two percent of older adolescent females were identified with IDA in our study, which was higher than the results found in 1994 in France (7.4% for adolescents aged from 14 to 18) [24,34]. The relevance of this comparison is somewhat questionable as there is a substantial time difference between the two studies: 2007 versus 1994. For adult females, it was estimated 1.7% (95% CI = 1.6%–1.8%) of IDA in France which was lower than results obtained in the SUpplementation en VItamines et Minéraux AntioXydants (SU.VI.MAX) study (4.4%) [36]. However, the latter study only included individuals aged from 35 to 60 years old.

To go further than the estimation of the prevalence and the number of cases, we have taken into account morbidity with the use of the composite metric DALY [37,38], which is the most used in risk and benefit assessment studies [30,39]. Here, only morbidity was considered, because there is no existing data about IDA fatalities in France. Three levels of IDA severity were taken into account: Mild, moderate, and severe. The sequelae (Table 2) of the different levels of IDA were assumed to be the same whatever the sub-population, and were based on the health status definitions from the GBD study [29]. This assumption may induce an underestimation of the IDA burden for children and adolescents. Indeed, GBD health status does not take into account possible disabilities such as neurodevelopment delay due to ID and IDA which may induce impaired learning abilities, memory skills, and behavioral issues [40,41]. The majority of the studies evaluating the beneficial effect of iron supplementation were carried out in children under three years old. However, this age is considered under our first age class. After this age, some studies found a beneficial effect of iron treatment in older anemic children on intelligence quotient, but not in non-anemic [42]. Moreover, there was no improvement in other disabilities, such as academic achievement, memory, or motor development [42]. Therefore, experts remain cautious about these conclusions and more studies are needed to conclude about this effect. The estimated DALY of IDA cases was 16 (95% CI = 11–20) per 100,000 people per year (Table 4) which corresponds to 9580 (95% CI = 7050–12,250) DALY for the French population per year. These values were lower than, but of the same order of magnitude as the estimation by GBD's study which reported 35 DALY per 100,000 French population in 2007 [43]. The major contributors to the burden were females aged from 25 to 44 and no longer adolescent girls aged from 15 to 17. This was due to the higher prevalence of moderate and severe anemia cases in the adult age class. However, our estimations were underestimated because children under three years old and pregnant females, which are particularly vulnerable populations, were excluded from the study [4,5].

The quantification of variability and uncertainties, as recommended by international organizations [44], took into account the fitted distribution of iron intake and disability weight. For the first input cited, uncertainty and variability were quantified separately. However, for the second input cited, this separation was not possible. Even if some classify it as only variability [45]—the confidence interval representing the variability of the severity of the disability—we decided to classify it as only uncertainty as done by the majority of the previous studies [46,47]. The variability of the iron needs and hemoglobin status were also taken into account in the study.

Nonetheless, despite the inherent uncertainty and variability described above, the results obtained in this study can be interpreted as showing that some differences in the burden of disease were observed, such as between males and females, or between menopausal and menstruating females.

The use of DALY metric also enabled to make consumption scenarios for concerned populations. Due to the richest source of heme-iron—which is the form of iron most absorbable by humans—brought by red meat consumption [11], ground beef was taken as a food item in intake scenarios. In addition, consumption of red meat was found to increase the level of serum ferritin in the blood [12,13]. It was estimated that for young children, 50 g/d of ground beef—corresponding to 1.3 mg of iron—added to the other sources of iron already present in current diet (without red meat), would be enough to almost eliminate the IDA burden. For females aged over 15 years old and adult males, the consumption of 100 g/d of ground beef—corresponding to 2.6 mg of iron—added to the other sources of iron already present in current diet (without red meat) would not be enough to eliminate IDA burden (Figure 2; Figure 3). These results were in line with French Pediatric Society recommendations of eating 100–150 g/d of meat for children under six years old who stopped increasing milk consumption and twice a day by children over seven years and adolescents [10], since beef meat, which has the highest level of heme-iron, is not the only meat consumed. The amount required to suppress IDA for adults is above the French Agency for Food, Environmental and Occupational Health and Safety (ANSES) recommendations, especially for adult women, which is to consume less than 500 g per week of red meat [48]. This recommendation was largely based on the conclusions of the World Cancer Research Found/Imperial College of London and the WHO, which classified red meat as "probably carcinogenic to humans" for colorectal cancer [49,50]. In addition, a recent study of our team determined that the consumption of over 65 g/d of red meat would increase the risk of cardiovascular disease for the most concerned populations (adult population, especially elderly) [51]. Indeed, heme-iron, is suspected to be the main component in the mechanism associated with colorectal cancer [52,53] and cardiovascular disease [54–56]. These latter health effects were not considered to be of concern for children and adolescents for whom current epidemiological data were available [57], but more studies and meta-analysis are needed to confirm this lack of association.

Furthermore, the intake scenario in this study does not consider the substitution and the acceptability of red meat. Indeed, when the individual increases the consumption of one food item, he/she may reduce the consumption of other foods. In addition, some population group will not accept to consume more red meat due to organoleptic reasons, beliefs, or ethics (e.g., vegetarian diet). In these cases, other sources of iron can be considered to reduce ID and IDA. The consumption of iron-rich sources such as white beans (7.97 mg of iron per 100 g), breakfast cereals (7.03 mg of iron per 100 g), or lentils (6.51 mg of iron per 100 g) [31] would be a possible solution, especially for adult populations with a risk of colorectal and cardiovascular disease. Nevertheless, the quantity consumed of these other sources of iron, which are in form of non heme-iron, should be higher than that needed for red meat, because the degree of absorption is lower than for heme-iron products. Otherwise, some other heme-iron containing products, such as fish and poultry have a lower proportion of iron in the food compared to red meat. Some studies show that the vegetarian and vegan population presented a higher risk of iron deficiencies [58–60]. However, there are large variations of non heme-iron absorption (0.7%–22.9%). Indeed, when blood levels had low serum and plasma ferritin concentrations, the absorption of iron increased [61]. In addition, dietary enhancers (e.g., ascorbic acid, alcohol) and inhibitors (e.g., calcium, tannins, and polyphenols) influence the absorption of non heme-iron [62].

Another possible solution for ID decrease is iron supplementation. It was observed that early iron supplementation of low birth weight infants would reduce the prevalence of behavioral problems and linear growth [3]. Another study showed that iron supplementation with vitamin C of IDA children aged from 6 to 12 in Turkey, increased the intelligence quotient by 4.8 points [63]. It was shown to improve learning and memory in non-anemic ID adolescent girls, in adults with IDA, and that iron supplementation can partially reverse the effects on cognitive performance.

5. Conclusions

The burden of the consumption of iron has been estimated and expressed in terms of prevalence, number of cases, and DALY. The study estimated 13,630 (95% CI = 5740–23,510) ID cases per 100,000 individuals per year. In terms of IDA, this corresponded to 1,290 (95% CI = 1230–1350) IDA cases per 100,000 people per year and to 16 (95% CI = 11–20) DALY per 100,000 individuals per year. In a further study, the IDA burden estimated here in terms of DALY will be balanced with other risks of iron-rich food intake, such as red meat. Indeed, for children and adolescents, the benefit of decreasing iron deficiency anemia can be compared with risks due to microbiological foodborne illnesses when eating red meat [64]. In addition to this latter risk, for adult populations, colorectal cancer and cardiovascular diseases attributable to red meat consumption [51,65] will be also considered in this overall risk–benefit assessment.

Author Contributions: Investigation, J.D.O.M.; conceptualization, J.M.M., F.P.; methodology, J.D.O.M., J.M.M., and P.T.; software, J.D.O.M. and J.M.M.; formal analysis, J.D.O.M.; validation, P.T.; supervision, F.P., J.M.M. and S.G.; writing—original draft, J.D.O.M.; writing—review & editing, J.D.O.M., J.M.M., F.P. and S.G.; funding acquisition, J.M.M.

Funding: This research was funded by Microbiologie et chaîne alimentaire department from the Institut National de la Recherche Agronomique and Région Pays de la Loire, grant number 34000653.

Acknowledgments: We acknowledge the "Equipe de Surveillance et d'Epidémiologie Nutritionnelle (Esen), équipe mixte de Santé publique France et de l'Université Paris 13", as the main investigator of the ENNS study and for the supply of results and the "Centres d'examen de santé de la Caisse nationale d'assurance maladie des travailleurs salariés (CnamTS) and their laboratories", for the determination of hemoglobin and ferritinemia.

Conflicts of Interest: The authors declare no conflict of interest. The funders had no role in the design of the study; in the collection, analyses, or interpretation of data; in the writing of the manuscript, or in the decision to publish the results.

Appendix A

Appendix A.1 Determination of the Hemoglobin Status in the Population

Table A1. Hemoglobin levels values of French population expressed in mean and (standard deviation) extracted from Stoltzfus et al. (2004) and Santé Publique France 2006–2007.

Age Class	Gender (and Status)	Hemoglobin (g/dL)	Reference
0–4	Male Female	12.5 (1) 12.5 (1)	[23]
5–14	Male Female	13.4 (1) 13.4 (1)	
15–29	Male Female	14.6 (1) 13.4 (1)	
18–24	Male Female (Premenopausal)	15.5 (0.8) 13.5 (1.5)	[26]
25–44	Male Female (Premenopausal) Female (Postmenopausal)	15.3 (1) 13.5 (1.4) 13.8 (0.6)	
45–64	Male Female (Premenopausal) Female (Postmenopausal)	15.3 (1.1) 13.7 (1.1) 13.8 (0.9)	
65–74	Male	14.9 (1.9)	
	Female (Postmenopausal)	14 (1.2)	

The levels of hemoglobin for the age classes under 18 years old (Table A1), did not correspond to the age classes used in this study. To determine hemoglobin status for the age classes 3–6, 7–11, 12–17 for male and 12–14 and 15–17 for females, the values were determined from linear functions, established from Stoltzfus et al. (2004) hemoglobin levels from Table A1.

Two linear functions were assumed to determine hemoglobin levels for females and one linear function for males aged under 18 years old:

- To determine female hemoglobin level in the age class three to six years old, the linear function was expressed:

$$Hb_{a,g} = 0.1173 \times mean\ age\ of\ the\ class + 12.275$$

- For females, age class from 7 to 11 and from 12 to 17 years old, the average hemoglobin levels were similar and estimated by the fallow function:

$$Hb_{a,g} = -0.0015 \times mean\ age\ of\ the\ class + 13.413$$

- To determine male hemoglobin level in the age class 3–6, 7–11, and 12–17 years old, the linear function was expressed:

$$Hb_{a,g} = 0.1058 \times mean\ age\ of\ the\ class + 12.331 \quad \text{(A1)}$$

Hemoglobin distribution for each age class was assumed to follow a normal distribution. To determine the mean standard deviation for age classes under 18 years old, the standard deviation of hemoglobin considered in this study was the same as used by Stoltzfus et al. (2004) for countries with anemia prevalence under 15% (1.0 g/dL) [9,23]. For older age classes, the standard deviation was estimated from ENNS study mean and confidence interval.

The levels of hemoglobin used in the present study is given in Table A2.

Table A2. Hemoglobin levels of the French population expressed in mean and (standard deviation) used in the present study.

Age Class	Gender (and Status)	Hemoglobin (g/dL)
3–6	Male Female	12.8 (1) 12.7 (1)
7–11	Male Female	13.3 (1) 13.4 (1)
12–17 12–14 15–17	Male Female Female (Premenopausal)	13.9 (1) 13.4 (1) 13.4 (1)
18–24	Male Female (Premenopausal)	15.5 (0.8) 13.5 (1.5)
25–44	Male Female (Premenopausal) Female (Postmenopausal)	15.3 (1) 13.5 (1.4) 13.8 (0.6)
45–64	Male Female (Premenopausal) Female (Postmenopausal)	15.3 (1.1) 13.7 (1.1) 13.8 (0.9)
65–74	Male	14.9 (1.9)
	Female (Postmenopausal)	14 (1.2)

Appendix A.2 Determining the Proportion of Anemia per Severity

To determine the proportion of anemia (*Ane*) per age class (*a*), gender (*g*), and severity (s), pnorm function was used from R software, considering hemoglobin distribution per age class and gender (*Hb*) in the French population (Table A2) and hemoglobin thresholds defined by WHO (Table 1) to define anemia per severity (*Hb.t*).

- Proportion of severe anemias:

$$Ane_{a,g,s=severe} = pnorm\left[Hb.t_{s=severe},\ mean\left(Hb_{a,g}\right),\ standard\ deviation\left(Hb_{a,g}\right)\right]$$

- Proportion of moderate anemias:

$$Ane_{a,g,s=moderate} = pnorm\left[Hb.t_{s=moderate},\ mean\left(Hb_{a,g}\right),\ standard\ deviation\left(Hb_{a,g}\right)\right] - Ane_{a,g,s=severe}$$

- Proportion of mild anemias:

$$Ane_{a,g,s=mild} = pnorm\left[Hb.t_{s=mild}, \ mean\left(Hb_{a,g}\right), \ standard\ deviation\left(Hb_{a,g}\right)\right] \\ -\left(Ane_{a,g,s=severe} + Ane_{a,g,s=moderate}\right)$$

- Allocation of anemia severity, among the anemia population

$$Alloc.Ane_{a,g,s} = \frac{Ane_{a,g,s}}{\sum Ane_{a,g,s}}$$

Table A3. Quantities of iron in different types of red meat.

Red Meat Type	Iron (mg/100 g)
Cooked bourguignon beef	4.3
Cooked stew beef	4.3
Roasted horse meat	3.4
Roasted beef	3.1
Cooked ground beef 5% fat	2.83
Cooked ground beef 10% fat	2.71
Braised beef	5.9
Raw ground beef 5% fat	2.65
Cooked ground beef 15% fat	2.6
Grilled entrecote beef	2.55
Cooked ground beef 20% fat	2.48
Roast beef ribbon	2.4
Burgundy fondue	2.3
Beef grilled steak	2.21
Roasted lamb shoulder	2.2
Cooked dumplings beef	2.2
Beef carpaccio	1.89
Unspecified cooked meat	1.77
Mixed skewer of meat	1.77
Other meat	1.77
Lamb skewer	1.73
Beef skewer	1.41
Roast lamb	1.4
Cooked veal side	1.3
Roasted lean pork tenderloin	1.29
Grilled lamb chop	1.27
Braised ribs	1.2
Breaded veal cutlet	1.11
Roasted loin pork	1.09
Cooked escalope veal	1
Simmered veal	0.95
Roasted tenderloin veal	0.87
Roasted veal	0.87
Grilled pork chop	0.84
Roasted pork	0.64

References

1. WHO/CDC. *Assessing the Iron Status of Populations: Including Literature Reviews*; WHO/CDC: Geneva, Switzerland, 2004; p. 112.
2. WHO/UNICEF/UNU. *Iron Deficiency Anaemia: Assessment, Prevention and Control, a Guide for Programme Managers*; World Health Organization: Geneva, Switzerland, 2001.

3. Jáuregui-Lobera, I. Iron deficiency and cognitive functions. *Neuropsychiatr. Dis. Treat.* **2014**, *10*, 2087–2095. [CrossRef] [PubMed]
4. Kassebaum, N.J.; on behalf of GBD 2013 Anemia Collaborators. The global burden of anemia. *Hematol. Oncol. Clin. N. Am.* **2016**, *30*, 247–308. [CrossRef] [PubMed]
5. WHO/CDC. *Worldwide Prevalence of Anaemia 1993–2005: WHO Global Database on Anaemia*; WHO/CDC: Geneva, Switzerland, 2008; p. 51.
6. Olives, J.P. Causes des déficits en fer chez l'enfant. *Arch. Pédiatrie* **2017**, *24*, 5S2–5S5. [CrossRef]
7. Dignass, A.; Farrag, K.; Stein, J. Limitations of Serum Ferritin in Diagnosing Iron Deficiency in Inflammatory Conditions. *Int. J. Chronic Dis.* **2018**, *2018*, 1–11. [CrossRef] [PubMed]
8. Soppi, E.T. Iron deficiency without anemia—A clinical challenge. *Clin. Case Rep.* **2018**, *6*, 1082–1086. [CrossRef] [PubMed]
9. Stoltzfus, R.J. Iron deficiency: Global prevalence and consequences. *Food Nutr. Bull.* **2003**, *24*, S99–S103. [CrossRef] [PubMed]
10. Tounian, P.; Chouraqui, J.P. Iron in nutrition. *Arch. Pediatr.* **2017**, *24*, 5s23–25s31. [CrossRef]
11. Czerwonka, M.; Tokarz, A. Iron in red meat–friend or foe. *Meat Sci.* **2017**, *123*, 157–165. [CrossRef] [PubMed]
12. Moshe, G.; Amitai, Y.; Korchia, G.; Korchia, L.; Tenenbaum, A.; Rosenblum, J.; Schechter, A. Anemia and iron deficiency in children: Association with red meat and poultry consumption. *J. Pediatr. Gastroenterol. Nutr.* **2013**, *57*, 722–727. [CrossRef]
13. Szymlek-Gay, A.E.; Ferguson, E.L.; Heath, A.L.M.; Gray, A.R.; Gibson, R.S.; Heath, A.L.M. Food-based strategies improve iron status in toddlers: A randomized controlled trial. *Am. J. Clin. Nutr.* **2009**, *90*, 1541–1551. [CrossRef]
14. WHO. Health Statistics and Information Systems: About the Global Burden of Disease (GBD) Project. Available online: http://www.who.int/healthinfo/global_burden_disease/about/en/ (accessed on 5 April 2018).
15. Cummins, E. Fundamental Principles of Risk Assessment. In *Textbook on Quantitative Tools for Sustainable Food and Energy in the Food Chain*; Geril, P., Ed.; EUROSIS: Ghent, Belgium, 2017; Volume 1, pp. 151–179.
16. Thompson, K.M. Variability and Uncertainty Meet Risk Management and Risk Communication. *Risk Anal.* **2002**, *22*, 647–654. [CrossRef] [PubMed]
17. Cummins, E. Quantifying Microbial Propagation. In *Modeling in Food Microbiology*; Membré, J.M., Vasilis, V., Eds.; Elsevier: Oxford, UK, 2016; pp. 17–31.
18. ANSES. Données de Consommations et Habitudes Alimentaires de L'étude INCA 2. Available online: https://www.data.gouv.fr/fr/datasets/donnees-de-consommations-et-habitudes-alimentaires-de-letude-inca-2-3/#_ (accessed on 23 August 2017).
19. Inserm. Ménopause: Améliorer la Sécurité D'utilisation des Traitements Hormonaux. Available online: https://www.inserm.fr/information-en-sante/dossiers-information/menopause (accessed on 1 June 2019).
20. EFSA Panel on Dietetic Products Nutrition Allergies. Scientific opinion on dietary reference values for iron. *EFSA J.* **2015**, *13*, 4254. [CrossRef]
21. Norden. *Nordic Nutrition Recommendations 2012: Integrating Nutrition and Physical Activity*; Nordic Council of Ministers: Copenhagen, Denmark, 2014; p. 627.
22. Ined. L'âge aux Premières Règles. Available online: https://www.ined.fr/fr/tout-savoir-population/memos-demo/focus/l-age-aux-premieres-regles/ (accessed on 20 June 2019).
23. Stoltzfus, R.J.; Mullany, L.; Black, R.E. *Iron Deficiency Anaemia*; WHO: Geneva, Switzerland, 2004; pp. 163–209.
24. Hercberg, S.; Preziosi, P.; Galan, P. Iron deficiency in Europe. *Public Health Nutr.* **2001**, *4*, 537–545. [CrossRef] [PubMed]
25. Hercberg, S.; Cailhol, J.; Franchisseur, C.; Maurel, M. La dificience en fer et l'anémie ferriprive dans la population française. *Rev. Française Lab.* **2001**, *334*, 55–58. [CrossRef]
26. Santé Publique France-Université Paris 13. *ENNS: Étude Nationale Nutrition Santé*; Santé Publique France: Saint-Maurice, France, 2006–2007.
27. Ferrari, M.; Mistura, L.; Patterson, E.; Sjöström, M.; E Díaz, L.; Stehle, P.; Gonzalez-Gross, M.; Kersting, M.; Widhalm, K.; Molnár, D.; et al. Evaluation of iron status in European adolescents through biochemical iron indicators: The HELENA Study. *Eur. J. Clin. Nutr.* **2011**, *65*, 340–349. [CrossRef] [PubMed]
28. Insee. *Projections de population 2007-2060 pour la France métropolitaine: Projections de population–Insee Résultats*; Insee: Paris, France, 2010.

29. Salomon, A.J.; Haagsma, A.J.; Davis, A.; De Noordhout, C.M.; Polinder, S.; Havelaar, A.H.; Cassini, A.; Devleesschauwer, B.; Kretzschmar, M.; Speybroeck, N.; et al. Disability weights for the Global Burden of Disease 2013 study. *Lancet Glob. Health* **2015**, *3*, e712–e723. [CrossRef]
30. Hoekstra, J.; Hart, A.; Boobis, A.; Claupein, E.; Cockburn, A.; Hunt, A.; Knudsen, I.; Richardson, D.; Schilter, B.; Schütte, K.; et al. BRAFO tiered approach for benefit–risk assessment of foods. *Food Chem. Toxicol.* **2012**, *50*, S684–S698. [CrossRef]
31. ANSES. Ciqual: Table de Composition Nutritionelle des Aliments. Available online: https://ciqual.anses.fr/ (accessed on 29 May 2019).
32. Dupont, C. Prévalence de la carence en fer. *Arch. Pédiatrie* **2017**, *24*, 5S45–5S48. [CrossRef]
33. Sacri, A.S.; Hercberg, S.; Gouya, L.; Levy, C.; Bocquet, A.; Blondel, B.; Vincelet, C.; Hebel, P.; Vinatier, I.; Montalembert, M.; et al. Very low prevalence of iron deficiency among young French children: A national cross-sectional hospital-based survey. *Matern. Child Nutr.* **2018**, *14*, e12460. [CrossRef]
34. Preziosi, P.; Hercberg, S.; Galan, P.; Devanlay, M.; Cherouvrier, F.; Dupin, H. Iron Status of a Healthy French Population: Factors Determining Biochemical Markers. *Ann. Nutr. Metab.* **1994**, *38*, 192–202. [CrossRef]
35. InVS. Marqueurs Biologiques de L'état Nutritionnel—Adultes 18–74 Ans: Tableaux de Distribution—Etude Nationale Nutrition Santé (ENNS). Available online: https://www.santepubliquefrance.fr/content/download/56036/file/etat-nutritionnel-adultes-vitamines-12.pdf (accessed on 30 August 2019).
36. Galan, P.; Yoon, H.C.; Preziosi, P.; Viteri, F.; Valeix, P.; Fieux, B.; Briançon, S.; Malvy, D.; Roussel, A.M.; Favier, A.; et al. Determining factors in the iron status of adult women in the SU.VI.MAX study. SUpplementation en VItamines et Minéraux AntioXydants. *Eur. J. Clin. Nutr.* **1998**, *52*, 383–388. [CrossRef] [PubMed]
37. Pires, S.M.; Boué, G.; Boobis, A.; Eneroth, H.; Hoekstra, J.; Membré, J.M.; Persson, I.M.; Poulsen, M.; Ruzante, J.; Van Klaveren, J.; et al. Risk Benefit Assessment of foods: Key findings from an international workshop. *Food Res. Int.* **2019**, *116*, 859–869. [CrossRef] [PubMed]
38. Murray, C.J. Quantifying the burden of disease: The technical basis for disability-adjusted life years. *Bull. World Health Organ.* **1994**, *72*, 429–445. [PubMed]
39. Nauta, M.J.; Andersen, R.; Pilegaard, K.; Pires, S.M.; Ravn-Haren, G.; Tetens, I.; Poulsen, M. Meeting the challenges in the development of risk-benefit assessment of foods. *Trends Food Sci. Technol.* **2018**, *76*, 90–100. [CrossRef]
40. Vallee, L. Iron and Neurodevelopment. *Arch. Pediatr.* **2017**, *24*, 5S18–5S22.
41. Tseng, M. Dietary Patterns and Prostate Cancer Risk in the National Health and Nutrition Examination Survey Epidemiological Follow-up Study Cohort. *Cancer Epidemiol. Biomark. Prev.* **2004**, *13*, 71–77. [CrossRef]
42. Larson, L.M.; Phiri, K.S.; Pasricha, S.-R. Iron and Cognitive Development: What Is the Evidence? *Ann. Nutr. Metab.* **2017**, *71*, 25–38. [CrossRef]
43. IHME. GBD Compare/Viz Hub. Available online: https://vizhub.healthdata.org/gbd-compare/ (accessed on 22 April 2019).
44. FAO/WHO; Codex Alimentarius Commission. *Principles and Guidelines for the Conduct of Microbiological Risk Assessment*; FAO: Rome, Italy, 1999; pp. 1–5.
45. Havelaar, A.H.; Van Duynhoven, Y.T.H.P.; Nauta, M.; Bouwknegt, M.; Heuvelink, A.E.; De Wit, G.A.; Nieuwenhuizen, M.G.M.; Van De Kar, N.C.A.J. Disease burden in The Netherlands due to infections with Shiga toxin-producing Escherichia coli O157. *Epidemiol. Infect.* **2004**, *132*, 467–484. [CrossRef]
46. Boué, G.; Cummins, E.; Guillou, S.; Antignac, J.; Le Bizec, B.; Membré, J. Development and Application of a Probabilistic Risk-Benefit Assessment Model for Infant Feeding Integrating Microbiological, Nutritional, and Chemical Components. *Risk Anal.* **2017**, *37*, 2360–2388. [CrossRef]
47. Thomsen, S.T.; Pires, S.M.; Devleesschauwer, B.; Poulsen, M.; Fagt, S.; Ygil, K.H.; Andersen, R. Investigating the risk-benefit balance of substituting red and processed meat with fish in a Danish diet. *Food Chem. Toxicol.* **2018**, *120*, 50–63. [CrossRef]
48. ANSES. *Avis et rapport Relatifs à L'actualisation des Repères du PNNS: Révision des Repères de Consommation Alimentaires*; ANSES: Maisons-Alfort, France, 2016; pp. 1–192.
49. WCRF/AICR. *Diet, Nutrition, Physical Activity and Colorectal Cancer*; WCRF/AICR: Washington, DC, USA, 2017; pp. 1–109.
50. Bouvard, V.; Loomis, D.; Guyton, K.Z.; Grosse, Y.; El Ghissassi, F.; Benbrahim-Tallaa, L.; Guha, N.; Mattock, H.; Straif, K. Carcinogenicity of consumption of red and processed meat. *Lancet Oncol.* **2015**, *16*, 1599–1600. [CrossRef]

51. Mota, J.D.O.; Boué, G.; Guillou, S.; Pierre, F.; Membré, J.M. Estimation of the burden of disease attributable to red meat consumption in France: Influence on colorectal cancer and cardiovascular diseases. *Food Chem. Toxicol.* **2019**, *130*, 174–186. [CrossRef] [PubMed]
52. Bastide, N.; Morois, S.; Cadeau, C.; Kangas, S.; Serafini, M.; Gusto, G.; Dossus, L.; Pierre, F.H.; Clavel-Chapelon, F.; Boutron-Ruault, M.C. Heme iron intake, dietary antioxidant capacity, and risk of colorectal adenomas in a large cohort study of French women. *Cancer Epidemiol. Biomark. Prev.* **2016**, *25*, 640–647. [CrossRef] [PubMed]
53. Bastide, N.M.; Chenni, F.; Audebert, M.; Santarelli, R.L.; Naud, N.; Baradat, M.; Jouanin, I.; Surya, R.; Hobbs, D.A.; Kuhnle, G.; et al. A Central Role for Heme Iron in Colon Carcinogenesis Associated with Red Meat Intake. *Cancer Res.* **2015**, *75*, 870–879. [CrossRef] [PubMed]
54. Ascherio, A.; Willett, W.C.; Rimm, E.B.; Giovannucci, E.L.; Stampfer, M.J. Dietary iron intake and risk of coronary disease among men. *Circulation* **1994**, *89*, 969–974. [CrossRef] [PubMed]
55. Qi, L.; van Dam, R.M.; Rexrode, K.; Hu, F.B. Heme iron from diet as a risk factor for coronary heart disease in women with type 2 diabetes. *Diabetes Care* **2007**, *30*, 101–106. [CrossRef] [PubMed]
56. Sullivan, J. Iron and the sex difference in heart disease risk. *Lancet* **1981**, *317*, 1293–1294. [CrossRef]
57. ANSES. *Avis de L'AGENCE Nationale de Sécurité Sanitaire de L'alimentation, de L'environnement et du Travail Relatif à L'actualisation des Repères Alimentaires du PNNS Pour les Enfants de 4 à 17 Ans*; ANSES: Maisons-Alfort, France, 2019; pp. 1–41.
58. Nelson, M.; White, J.; Rhodes, C. Haemoglobin, ferritin, and iron intakes in British children aged 12–14 years: A preliminary investigation. *Br. J. Nutr.* **1993**, *70*, 147–155. [CrossRef]
59. Donovan, U.M.; Gibson, R.S. Iron and zinc status of young women aged 14 to 19 years consuming vegetarian and omnivorous diets. *J. Am. Coll. Nutr.* **1995**, *14*, 463–472. [CrossRef]
60. Alexander, D.; Ball, M.J.; Mann, J. Nutrient intake and haematological status of vegetarians and age-sex matched omnivores. *Eur. J. Clin. Nutr.* **1994**, *48*, 538–546.
61. Collings, R.; Harvey, L.J.; Hooper, L.; Hurst, R.; Brown, T.J.; Ansett, J.; King, M.; Fairweather-Tait, S.J. The absorption of iron from whole diets: A systematic review. *Am. J. Clin. Nutr.* **2013**, *98*, 65–81. [CrossRef] [PubMed]
62. Hurrell, R.; Egli, I. Iron bioavailability and dietary reference values. *Am. J. Clin. Nutr.* **2010**, *91*, 1461S–1467S. [CrossRef] [PubMed]
63. Agaoglu, L.; Torun, O.; Unuvar, E.; Sefil, Y.; Demir, D. Effects of iron deficiency anemia on cognitive function in children. *Arzneimittelforschung* **2007**, *57*, 426–430. [CrossRef] [PubMed]
64. ECDC/EFSA. *The European Union Summary Report on Trends and Sources of Zoonoses, Zoonotic Agents and Food-Borne Outbreaks in 2017*; European Food Safety Authority: Parma, Italy, 2018; p. 262.
65. Mota, J.D.O.; Tounian, P.; Guillou, S.; Pierre, F.; Membré, J.-M. Risque et bénéfice nutritionnels associés à la consommation de viande rouge en France. *Nutr. Clin. Métabolisme* **2019**, *33*, 109. [CrossRef]

nutrients

MDPI

Article

Total and Nonheme Dietary Iron Intake Is Associated with Metabolic Syndrome and Its Components in Chinese Men and Women

Zhenni Zhu [1,2], Fan Wu [2], Ye Lu [3], Chunfeng Wu [4], Zhengyuan Wang [2], Jiajie Zang [2], Changyi Guo [2], Xiaodong Jia [2], Jiahui Yao [5], Hui Peng [6], Yuna He [1], Jing Sun [1], Jian Huang [1] and Gangqiang Ding [1,*]

1 National Institute for Nutrition and Health, Chinese Center for Disease Control and Prevention, 27 Nanwei Road, Xicheng District, Beijing 100050, China; zhuzhenni@scdc.sh.cn (Z.Z.); heyn@ninh.chinacdc.cn (Y.H.); sunjing@ninh.chinacdc.cn (J.S.); huangjian@ninh.chinacdc.cn (J.H.)

2 Division of Health Risk Factors Monitoring and Control, Shanghai Municipal Center for Disease Control and Prevention, 1380 West Zhongshan Road, Shanghai 20036, China; wufan@scdc.sh.cn (F.W.); wangzhengyuan@scdc.sh.cn (Z.W.); zangjiajie@scdc.sh.cn (J.Z.); guochangyi@scdc.sh.cn (C.G.); jiaxiaodong@scdc.sh.cn (X.J.)

3 Division of Non-Communicable Diseases Prevention and Control, Shanghai Municipal Center for Disease Control and Prevention, 1380 West Zhongshan Road, Shanghai 20036, China; luye@scdc.sh.cn

4 Department of Profession Management, Shanghai Municipal Center for Disease Control and Prevention, 1380 West Zhongshan Road, Shanghai 20036, China; wuchunfeng@scdc.sh.cn

5 Department of Public Health, Shanghai Xuhui District Center for Disease Control and Prevention, 50 Yongchuan Road, Shanghai 200237, China; jeffkokeyjh@126.com

6 Department of Public Health, Shanghai Jiading District Center for Disease Control and Prevention, 264 Tacheng Road, Shanghai 201800, China; jzhjymzl@126.com

* Correspondence: dinggq@chinacdc.cn; Tel.: +86-10-6623-7001

Received: 13 October 2018; Accepted: 31 October 2018; Published: 4 November 2018

Abstract: The causal relationship between serum ferritin and metabolic syndrome (MetS) remains inconclusive. Dietary iron intake increases serum ferritin. The objective of this study was to evaluate associations of total, heme, and nonheme dietary iron intake with MetS and its components in men and women in metropolitan China. Data from 3099 participants in the Shanghai Diet and Health Survey (SDHS) obtained during 2012–2013 were included in this analysis. Dietary intake was assessed by 24-h diet records from 3 consecutive days. Multivariate generalized linear mixed models were used to evaluate the associations of dietary iron intake with MetS and its components. After adjustment for potential confounders as age, sex, income, physical exercise, smoking status, alcohol use, and energy intake, a positive trend was observed across quartiles of total iron intake and risk of MetS (p for trend = 0.022). Compared with the lowest quartile of total iron intake (<12.72 mg/day), the highest quartile ($\geq$21.88 mg/day) had an odds ratio (95% confidence interval), OR (95% CI), of 1.59 (1.15,2.20). In addition, the highest quartile of nonheme iron intake ($\geq$20.10 mg/day) had a 1.44-fold higher risk of MetS compared with the lowest quartile (<11.62 mg/day), and higher risks of MetS components were associated with the third quartiles of total and nonheme iron intake. There was no association between heme iron intake and risk of MetS (p for trend = 0.895). Associations for total and nonheme iron intake with MetS risk were found in men but not in women. Total and nonheme dietary iron intake was found to be positively associated with MetS and its components in the adult population in metropolitan China. This research also revealed a gender difference in the association between dietary iron intake and MetS.

Keywords: dietary iron intake; heme iron; nonheme iron; metabolic syndrome; population-based study

1. Introduction

China has experienced a cardiovascular disease epidemic in recent decades [1]. Metabolic syndrome (MetS) refers to a constellation of interrelated risk factors that increase the development of both cardiovascular disease and type 2 diabetes mellitus [2]. The current nationwide prevalence of MetS among Chinese adults is 24.2%, which is an astonishing increase compared with the value of 9.8% calculated one decade ago based on the same diagnostic criteria [3,4].

A few studies have demonstrated an association between serum ferritin and MetS [5–8], but whether MetS is causally related to higher levels of iron metabolic markers or whether higher levels of iron metabolic markers lead to MetS remains inconclusive. Iron overload is characterized physiologically by an increase in the serum ferritin levels [9], and some studies have shown that meat or heme iron intake is related to the ferritin levels in serum [10,11]. However, high levels of serum ferritin can also be produced as a consequence of metabolic disorders, regardless of the iron overload status [12].

Few studies have investigated the relationship between dietary iron intake and MetS, which could potentially address the causative nature of the association between iron metabolic markers and MetS. Iron is present in foods in a heme or nonheme form. Heme iron is in most animal foods, and the rest of the iron in animal or plant food is nonheme iron [13]. Heme iron is more efficiently absorbed than nonheme iron as approximately 25% heme iron and 5% nonheme iron from diet absorbed by body [14]. The Chinese diet is known to be plant-based, which implies a low bioavailability of dietary iron. The primary objective of this study was to evaluate the associations of dietary iron intake, in terms of total, heme and nonheme iron intake, with MetS and its components in the adult population in metropolitan China. Considering the different physiologic mechanisms underlying body iron loss among men and women, it was intended to also analyze these associations stratified by gender. To the best of our knowledge, this population-based study constitutes the first investigation of the relationship between dietary iron intake and MetS in a Chinese population.

2. Materials and Methods

2.1. Study Population

The data used in this study were obtained from a cross-sectional investigation, the first wave of the Shanghai Diet and Health Survey (SDHS), which was conducted in Shanghai, one of the most developed cities in China, during the period 2012–2013. The SDHS was designed to examine the associations of food consumption, energy and nutrient intake, and behavioral factors with nutrition-related health outcomes among local residents. The investigation enrolled 4504 community-dwelling men and women aged at least 18 years from 54 randomly stratified-sampled communities. The participants were initially selected to constitute a random representative sample of the local adult population (n = 1725), and their family members aged at least 18 years were then recruited (approximately 2.6 persons per family). Those who had lived in the area for less than 6 months in total during the last year of the survey were excluded. The Shanghai Municipal Center for Disease Control and Prevention was responsible for the implementation of the SDHS.

Participants with missing anthropometric measurements (n = 251) or lack of blood pressure assessed (n = 54) or whose blood samples were not collected or tested (n = 817) for indicators of MetS diagnosis, as well as those who reported an energy intake less than 300 kcal/d or greater than 3500 kcal/d for the average energy intake (mean ± SD) was 1766 ± 880 kcal in the study population and the mean ± 2SD interval was finally used for the acceptable range (n = 51), were excluded. Participants with missing dietary survey information (n = 38) or other pertinent covariates (n = 194) were also excluded. The data from 3099 participants were ultimately included in the present analysis.

The SDHS was approved by the Shanghai Municipal Center for Disease Control and Prevention's Institutional Review Board on 31 January 2012. Informed consent was obtained from each participant before the survey. The study complied with the code of ethics of the World Medical Association (Declaration of Helsinki).

2.2. Dietary Assessment

The dietary survey included a 24-h dietary record for 3 consecutive days (including 2 weekdays and 1 weekend day) in the consideration of different dietary behavior between working days and rest, as well as the feasibility of investigation. Household condiments mainly containing fat or sodium, including cooking oil, salt, soy sauce, chili sauce, etc., were weighed before and after the 3 survey days. The interviewers were public health doctors from 54 local community health centers who received a standard training course on the recording of dietary information. Each participant was orally instructed to record their daily food intake both at home and out of home on draft paper at the beginning and interviewed face-to-face by interviewers in the consecutive survey days at home. At each survey day, the interviewers collected and checked though the draft paper, and afterward, revised the food weight and transcribed the draft dietary information into a structured form. Furthermore, the participants were instructed not to change their typical diet or physical activity during the survey period. The diet records in the structured form were reviewed by nutrition specialists from local centers of disease control and prevention. No disastrous events such as rain or snow disasters that would have affected the normal food supply occurred during the survey period.

Daily food consumption was calculated from the 3-day, 24-h diet record. The 3-day consumption of condiments based on the calculated weight difference was divided into individual intake according to the eating times in the home and the individuals' energy intake (only from food) proportion among family members. That consumption of condiments from meals out of home was simulated according to the previously calculated condiments' densities in food consumed at home. The intake of dietary energy, macronutrients and iron was estimated according to daily food and condiment consumption using the Chinese food composition database [15,16]. Dietary supplements and medications were excluded from nutrient intake. Heme iron was estimated as 40% of the total iron from animal foods, including red meat, poultry, fish, and animal organs, and nonheme iron was calculated as the remaining portion of the total iron from all foods [14,17].

2.3. Potential Confounders

Information on each participant's age, sex, education, income, smoking status, alcohol use, physical activity level, and intentional physical exercise was recorded using an interviewer-administered questionnaire at each participant's home. The educational level of the participants was reported as years of education. The yearly income was calculated by dividing the total family yearly income by the number of family members. The physical activity level was recorded as sedentary, moderate, or vigorous according to professional and nonprofessional activities. Intentional physical exercise was defined as physical exercise performed for the purpose of health maintenance or fitness. The smoking status was categorized as never smoked, former smoker, or current smoker. With respect to alcohol use, the respondents were classified as follows according to their alcohol consumption during the seven days prior to the interview: lifetime abstainers, the respondent did not consume an alcoholic beverage; nonheavy drinkers (social drinkers), the respondent consumed 5+ standard drinks once (1 day) in the 1-week period; infrequent heavy drinkers (binge drinkers), the respondent consumed 5+ standard drinks on 2–3 days in the 1-week period; or frequent heavy drinkers, the respondent consumed 5+ standard drinks on at least 4 days in the 1-week period.

2.4. Anthropometric and Laboratory Measurements

All anthropometric measurements were conducted in the community health centers located in each participant's community. The waist circumference was measured using a Graham-Field 1340-2 tape measure. The blood pressure was measured three times after a quiet rest for 5 min using an Omron HEM-7071 electronic sphygmomanometer (Omron Healthcare, Kyoto, Japan).

Each participant was asked to fast for more than 10 h, and their blood was then collected and analyzed at the laboratory of the Shanghai Municipal Centers for Disease Control and Prevention.

The serum concentrations of glucose, triglycerides, and high-density lipoprotein-cholesterol (HDL-C) were measured using a HITACHI 7080 Automatic Biochemical Analyzer with reagents from Wako Pure Chemical Industries, Ltd. (Tokyo, Japan).

2.5. Definition of Metabolic Syndrome

MetS was identified based on the criteria in the US National Cholesterol Education Program Adult Treatment Panel III (NCEP-ATP III) for Asian populations [2], which states that at least three of the following metabolic abnormalities should be present: (1) elevated waist circumference (WC $\geq$ 90 cm for men and WC $\geq$ 80 cm for women); (2) elevated triglycerides (triglycerides $\geq$ 150 mg/dL) or on drug treatment for elevated triglycerides; (3) reduced HDL-C (HDL-C < 40 mg/dL for men and < 50 mg/dL for women) or on drug treatment for reduced HDL-C; (4) elevated blood pressure (systolic blood pressure $\geq$ 130 mmHg and/or diastolic blood pressure $\geq$ 85 mmHg) or on antihypertensive drug treatment with a history of hypertension; and (5) elevated fasting glucose (100 mg/dL) or on drug treatment for elevated glucose.

2.6. Statistical Analyses

Statistical analyses were conducted using SAS statistical software (v. 9.2; SAS Institute, Cary, NC, USA). Given that the occurrence of MetS might aggregate in families due to a similar genetic background, which would potentially cause aggregation bias, multilevel models were introduced in the analysis. Multivariate generalized linear mixed models for binary data with a logit link function were applied to determine the odds ratios (ORs) and 95% confidence intervals (CIs) of the occurrence of MetS and its five components as the dependent variables based on four tertiles of total, heme, and nonheme dietary iron intake as the independent fixed-effect variables and family aggregation as the random-effect variable. No covariate was included in the crude model. Potential confounders, including age, sex, income, physical activity level, intentional physical exercise, dietary energy intake, smoking status, and alcohol use, were introduced as covariates in the adjusted models. To examine the linear trend between dietary iron intake and the occurrence of MetS, the medians in each quartile of dietary iron intake were used as the independent fixed-effect variables, and the occurrences of MetS and its five components were used as the dependent variables. A two-sided p value < 0.05 was considered to indicate statistical significance.

3. Results

3.1. Characteristics of the Participants

The analysis sample included 3099 participants, consisting of 1430 male adults and 1669 female adults. Among the participants, the total, heme, and nonheme dietary iron intake was 19.7 mg/day, 1.6 mg/day, and 18.1 mg/day, respectively, and the prevalence of MetS was 23.9%. The characteristics of the participants are shown in Table 1.

3.2. Dietary Sources of Iron Intake

In present study population, the main dietary sources of total iron intake were plant-based foods. Specifically, the top three dietary sources of total iron intake were grains/potatoes, vegetables, and condiments, and this finding was obtained for both males (6.62 mg/day, 4.92 mg/day, and 2.28 mg/day, respectively) and females (5.10 mg/day, 4.29 mg/day, and 1.97 mg/day, respectively). Red meat was ranked first among the animal food sources of total dietary iron intake consumed by both genders (Figure 1).

3.3. Quartiles of Dietary Iron Intake and MetS Risk

After adjusting for age, sex, income, physical activity level, intentional physical exercise, dietary energy intake, smoking status, and alcohol use, a significant positive trend was found across quartiles of total dietary iron intake and risk of MetS (p for trend = 0.022). Compared with the lowest quartile of dietary total iron intake (<12.72 mg/day), the highest quartile (≥21.88 mg/day) had an odds ratio (95% confidence interval), OR (95% CI), of 1.59 (1.15,2.20). However, after adjusting for the same confounders, there was no association between heme dietary iron intake and risk of MetS (p for trend = 0.895), and compared with the lowest quartile of heme iron intake, the second to fourth tertiles did not show significant differences in ORs. Although the quartiles of nonheme iron intake were not in line with the risk of MetS (p for trend = 0.065), the highest quartile of nonheme iron intake (≥20.10 mg/day) had a 1.44-fold higher risk of MetS compared with the lowest quartile (<11.62 mg/day). After adjustment, higher risks for the five components of MetS were observed in the third and fourth quartiles of total and nonheme dietary iron intake. However, no differences in the risks for the MetS components were found between the second to fourth quartiles of heme iron intake and the reference quartile (Tables 2–4).

After stratifying by gender, the linear trend between the quartiles of total dietary iron intake and risk of MetS was still significant in male participants (p for trend = 0.016) but not in female participants (p for trend = 0.322). In men, a higher risk (OR = 2.11, 95% CI: 1.29–3.45) of MetS was observed in the fourth quartile of total dietary iron intake (≥23.68 mg/day) compared with the first quartile (<14.12 mg/day), whereas in women, the risks of MetS were not significantly different between the lowest quartile of total iron intake and the other quartiles. A positive association was found between nonheme iron intake and MetS risk in men (p for trend = 0.025) but not in women (p for trend = 0.150). Finally, heme iron intake was not associated with MetS in either men or women (Men, Tables 5–7; Women, Tables 8–10).

Table 1. Characteristics of the participants by gender in the SDHS 2012–2013.

	All	Male	Female	p-Value
n (%)	3099 (100.0)	1430 (46.1)	1669 (53.9)	
Age, %				0.513
18–44 years	30.8	30.1	31.4	
45–59 years	36.2	36.3	36.1	
60-years	33.0	33.6	32.6	
Yearly Income, %				0.523
Above average level (>60,000 RMB [1])	7.2	7.0	7.4	
Average level (30,000–59, RMB)	30.9	32.1	30.0	
Below average level (<30,000 RMB)	57.2	56.0	58.2	
No answer	4.7	4.9	4.5	
Years of Education, years (SD [2])	9.5 (4.53)	10.2 (4.01)	8.9 (4.86)	<0.001
Physical Activity Level, %				<0.001
Sedentary	84.2	78.7	89.0	
Moderate	13.8	18.1	10.2	
Vigorous	2.0	3.3	0.8	
Intentional Physical Exercise, %	24.8	25.2	24.5	0.940
Smoking Status, %				
Never smoked	72.0	40.8	98.8	
Former smoker	5.0	10.5	0.3	
Current smoker	23.0	48.8	1.0	
Alcohol Use, %				<0.001
Lifetime abstainers	80.7	64.0	94.9	
Nonheavy drinkers	15.1	27.3	4.6	
Infrequent heavy drinkers	1.3	2.6	0.3	
Frequent heavy drinkers	3.0	6.1	0.3	
Dietary Intake				
Energy, kcal/day (SD)	1760.9 (842.5)	1938.5 (884.5)	1608.9(773.4)	<0.001
Carbohydrate, g/day (SD)	207.4(119.3)	226.8(131.5)	190.8(105.1)	<0.001
Protein, g/day (SD)	67.9(45.6)	75.0(53.1)	61.9(37.1)	<0.001
Total fat, g/day (SD)	74.6(41.8)	81.1(40.7)	69.0(41.9)	<0.001
Total iron, mg/day (SD)	19.7(16.3)	22.0(20.4)	17.7(11.3)	<0.001
Heme iron, mg/day (SD)	1.6(1.4)	1.7(1.4)	1.5(1.4)	<0.001
Nonheme iron, mg/day (SD)	18.1(15.8)	20.3(20.0)	16.2(10.6)	<0.001
Metabolic Syndrome, %	23.9	21.8	25.7	0.011
Metabolic Syndrome Components				
Elevated blood pressure, %	52.4	56.8	48.7	<0.001
Elevated waist circumference, %	33.6	25.5	40.5	<0.001
Elevated fasting glucose, %	23.1	25.4	21.1	0.007
Elevated triglycerides, %	27.2	30.3	24.4	<0.001
Reduced HDL-C [3], %	21.0	14.8	26.2	<0.001

[1] RMB, renminbi, China's currency; [2] SD, standard deviation; [3] HDL-C, high-density lipoprotein-cholesterol.

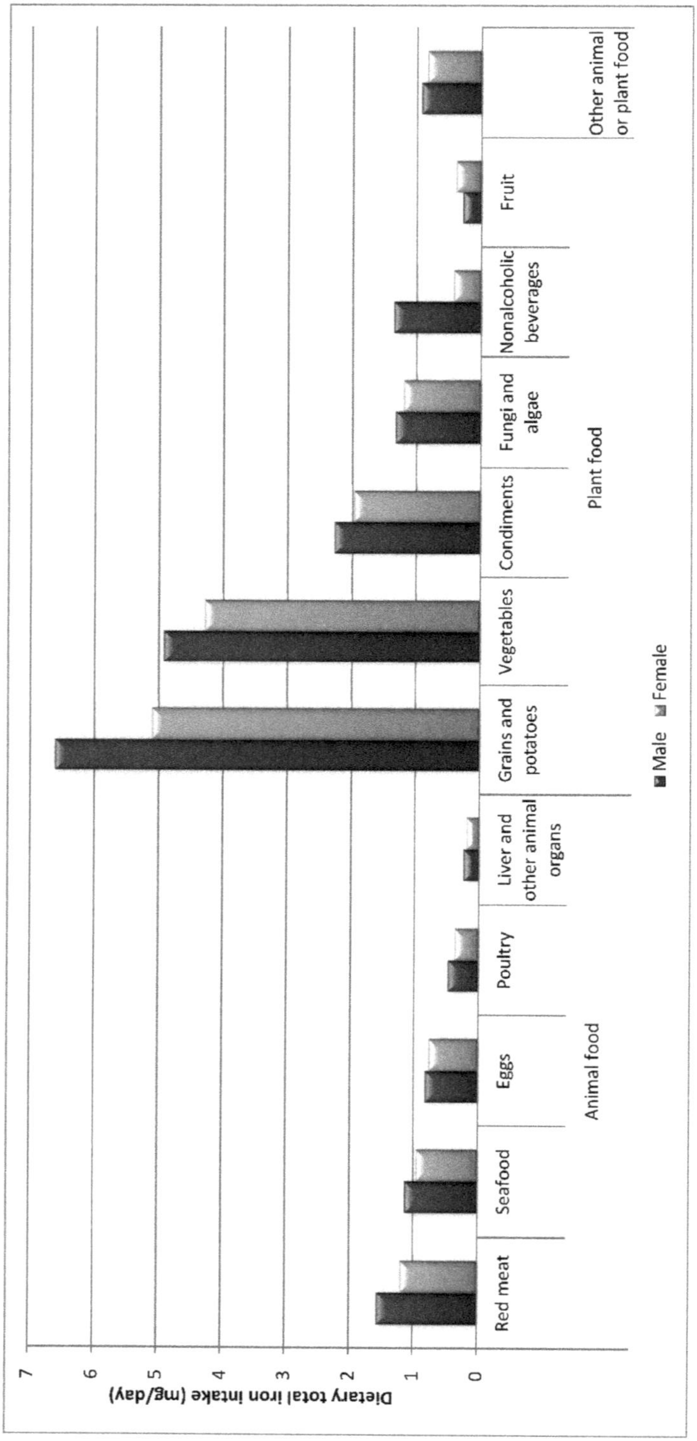

Figure 1. Dietary sources of total iron intake for the different genders in the SDHS 2012–2013.

Table 2. Odds ratios (ORs) (95% CI) for metabolic syndrome and its components according to the quartiles of total iron intake (mg/day) among all participants in the SDHS 2012–2013.

	Quartiles of Dietary Iron Intake, ORs (95% CI)				
	Q1	Q2	Q3	Q4	*p*-Value for Trend [1]
Total Iron Intake (mg/day)	<12.72	(12.72–16.50)	(16.50–21.88)	≥21.88	
n	776	773	776	774	
Metabolic syndrome					
Crude model	Reference	1.12(0.88,1.42)	1.24(0.98,1.57)	1.17(0.92,1.48)	0.335
Adjusted model [2]	Reference	1.37(1.06,1.78) *	1.47(1.11,1.94) **	1.59(1.15,2.20) **	0.022
				p-Value for interaction [3] = 0.060	
Metabolic syndrome components					
Elevated blood pressure					
Crude model	Reference	0.95(0.79,1.13)	1.19(1.00,1.42) *	1.07(0.89,1.27)	0.063
Adjusted model [2]	Reference	1.14(0.92,1.41)	1.33(1.06,1.66) *	1.21(0.93,1.57)	0.100
Elevated waist circumference					
Crude model	Reference	1.07(0.89,1.29)	1.09(0.91,1.32)	1.00(0.83,1.21)	0.691
Adjusted model [2]	Reference	1.25(1.01,1.53) *	1.29(1.04,1.60) *	1.25(0.96,1.61)	0.096
Elevated fasting glucose					
Crude model	Reference	1.03(0.83,1.27)	1.16(0.94,1.42)	1.12(0.91,1.38)	0.458
Adjusted model [2]	Reference	1.18(0.94,1.49)	1.25(0.98,1.60)	1.33(1.00,1.77)	0.221
Elevated triglycerides					
Crude model	Reference	1.17(0.95,1.43)	1.46(1.20,1.77) **	1.34(1.10,1.64) **	0.001
Adjusted model [2]	Reference	1.14(0.92,1.42)	1.37(1.09,1.72) **	1.31(1.00,1.71) *	0.047
Reduced HDL-C [4]					
Crude model	Reference	1.09(0.86,1.38)	1.11(0.87,1.41)	0.91(0.71,1.16)	0.348
Adjusted model [2]	Reference	1.27(0.98,1.65)	1.42(1.08,1.87) *	1.31(0.94,1.83)	0.088

* $p < 0.05$, ** $p < 0.001$. [1] The *p*-value for the trend was examined using the medians in each quartile of dietary iron intake. [2] The models were adjusted for age, sex, income, physical activity level, intentional physical exercise, smoking status, alcohol use and dietary total energy intake. [3] The *p*-value for the interaction was tested for the quartiles of dietary iron intake by gender. [4] HDL-C: high-density lipoprotein-cholesterol.

Table 3. ORs (95% CI) for metabolic syndrome and its components according to the quartiles of heme iron intake (mg/day) among all participants in the SDHS 2012–2013.

	Quartiles of Dietary Iron Intake, ORs (95% CI)				
	Q1	Q2	Q3	Q4	*p*-Value for Trend [1]
Heme Iron Intake (mg/day)	<0.83	(0.83–1.28)	(1.28–1.94)	≥1.94	
n	776	774	773	776	
Metabolic syndrome					
Crude model	Reference	0.91(0.72,1.15)	0.92(0.73,1.16)	0.85(0.67,1.07)	0.593
Adjusted model [2]	Reference	1.06(0.82,1.36)	1.11(0.86,1.44)	1.06(0.80,1.39)	0.895
				p-Value for interaction [3] = 0.096	
Metabolic syndrome components					
Elevated blood pressure					
Crude model	Reference	0.76(0.64,0.91) **	0.73(0.61,0.87) **	0.73(0.61,0.87) **	0.001
Adjusted model [2]	Reference	0.91(0.74,1.13)	0.98(0.79,1.22)	1.05(0.83,1.32)	0.620
Elevated waist circumference					
Crude model	Reference	0.89(0.74,1.08)	0.95(0.79,1.14)	0.93(0.77,1.12)	0.681
Adjusted model [2]	Reference	1.07(0.87,1.31)	1.20(0.98,1.48)	1.24(0.99,1.55)	0.189
Elevated fasting glucose					
Crude model	Reference	0.90(0.73,1.11)	0.82(0.67,1.01)	0.89(0.73,1.10)	0.330
Adjusted model [2]	Reference	1.02(0.82,1.28)	0.96(0.76,1.21)	1.11(0.87,1.41)	0.661
Elevated triglycerides					
Crude model	Reference	1.05(0.86,1.29)	1.12(0.92,1.36)	1.08(0.89,1.32)	0.723
Adjusted model [2]	Reference	1.03(0.83,1.28)	1.10(0.88,1.36)	1.00(0.79,1.26)	0.784
Reduced HDL-C [4]					
Crude model	Reference	1.05(0.83,1.35)	1.16(0.91,1.47)	0.99(0.78,1.27)	0.547
Adjusted model [2]	Reference	1.17(0.91,1.52)	1.29(0.99,1.68)	1.15(0.87,1.53)	0.291

* $p < 0.05$, ** $p < 0.001$. [1] The *p*-value for the trend was examined using the medians in each quartile of dietary iron intake. [2] The models were adjusted for age, sex, income, physical activity level, intentional physical exercise, smoking status, alcohol use and dietary total energy intake. [3] The *p*-value for the interaction was tested for the quartiles of dietary iron intake by gender. [4] HDL-C: high-density lipoprotein-cholesterol.

Table 4. ORs (95% CI) for metabolic syndrome and its components according to the quartiles of nonheme iron intake (mg/day) among all participants in the SDHS 2012–2013.

	Quartiles of Dietary Iron Intake, ORs (95% CI)				
	Q1	**Q2**	**Q3**	**Q4**	***p*-Value for Trend** [1]
Nonheme Iron Intake (mg/day)	<11.62	(11.62–15.10)	(15.10–20.10)	≥20.10	
n	774	774	775	776	
Metabolic syndrome					
Crude model	Reference	1.07(0.84,1.36)	1.23(0.97,1.55)	1.12(0.89,1.42)	0.364
Adjusted model [2]	Reference	1.28(0.98,1.66)	1.43(1.09,1.87) *	1.44(1.04,1.99) *	0.065
				p-Value for interaction [3] = 0.086	
Metabolic syndrome components					
Elevated blood pressure					
Crude model	Reference	0.91(0.77,1.09)	1.26(1.06,1.50) **	1.08(0.91,1.29)	0.003
Adjusted model [2]	Reference	1.07(0.86,1.32)	1.34(1.07,1.67) *	1.18(0.91,1.53)	0.058
Elevated waist circumference					
Crude model	Reference	1.07(0.89,1.29)	1.10(0.91,1.32)	1.00(0.83,1.20)	0.662
Adjusted model [2]	Reference	1.22(0.99,1.50)	1.26(1.02,1.57) *	1.19(0.92,1.53)	0.149
Elevated fasting glucose					
Crude model	Reference	0.92(0.74,1.14)	1.19(0.97,1.46)	1.08(0.88,1.33)	0.088
Adjusted model [2]	Reference	1.01(0.80,1.28)	1.28(1.01,1.63) *	1.25(0.94,1.66)	0.107
Elevated triglycerides					
Crude model	Reference	1.21(0.99,1.48)	1.42(1.17,1.73) **	1.34(1.10,1.64) **	0.003
Adjusted model [2]	Reference	1.15(0.93,1.44)	1.33(1.06,1.66) *	1.26(0.96,1.64)	0.100
Reduced HDL-C [4]					
Crude model	Reference	1.11(0.87,1.40)	1.06(0.83,1.35)	0.91(0.71,1.16)	0.425
Adjusted model [2]	Reference	1.29(0.99,1.67)	1.34(1.02,1.76) *	1.30(0.93,1.81)	0.157

* $p < 0.05$, ** $p < 0.001$. [1] The *p*-value for the trend was examined using the medians in each quartile of dietary iron intake. [2] The models were adjusted for age, sex, income, physical activity level, intentional physical exercise, smoking status, alcohol use, and dietary total energy intake. [3] The *p*-value for the interaction was tested for the quartiles of dietary iron intake by gender. [4] HDL-C: high-density lipoprotein-cholesterol.

Table 5. ORs (95% CI) for metabolic syndrome and its components according to the quartiles of total iron intake (mg/day) among men in the SDHS 2012–2013.

	Quartiles of Dietary Iron Intake, ORs (95% CI)				
	Q1	Q2	Q3	Q4	p-Value for Trend [1]
Total Iron Intake (mg/day)	<14.12	(14.12–17.87)	(17.87–23.68)	≥23.68	
n	357	358	358	357	
Metabolic syndrome					
Crude model	Reference	1.22(0.84,1.77)	1.66(1.16,2.39) **	1.65(1.15,2.36) **	0.013
Adjusted model [2]	Reference	1.36(0.90,2.07)	1.83(1.19,2.81) **	2.11(1.29,3.45) **	0.016
Metabolic syndrome components					
Elevated blood pressure					
Crude model	Reference	1.07(0.83,1.38)	1.15(0.90,1.48)	1.16(0.90,1.49)	0.619
Adjusted model [2]	Reference	1.35(1.00,1.84)	1.33(0.97,1.82)	1.43(0.99,2.06)	0.164
Elevated waist circumference					
Crude model	Reference	1.65(1.21,2.24) **	1.74(1.29,2.36) **	1.84(1.36,2.49) **	0.000
Adjusted model [2]	Reference	1.59(1.14,2.21) **	1.61(1.15,2.27) **	1.65(1.12,2.43) *	0.019
Elevated fasting glucose					
Crude model	Reference	0.94(0.70,1.26)	1.00(0.75,1.33)	0.98(0.74,1.31)	0.973
Adjusted model [2]	Reference	1.14(0.82,1.58)	1.17(0.83,1.66)	1.40(0.94,2.10)	0.431
Elevated triglycerides					
Crude model	Reference	1.24(0.94,1.64)	1.38(1.05,1.82) *	1.32(1.01,1.74) *	0.096
Adjusted model [2]	Reference	1.21(0.89,1.65)	1.30(0.95,1.78)	1.34(0.93,1.94)	0.361
Reduced HDL-C					
Crude model	Reference	0.98(0.65,1.48)	1.23(0.83,1.83)	1.08(0.72,1.62)	0.658
Adjusted model [2]	Reference	0.92(0.59,1.45)	1.38(0.87,2.19)	1.34(0.77,2.32)	0.287

* $p < 0.05$, ** $p < 0.001$. [1] The p-value for the trend was examined using the medians in each quartile of dietary iron intake. [2] The models were adjusted for age, income, physical activity level, intentional physical exercise, smoking status, alcohol use, and dietary total energy intake.

Table 6. ORs (95% CI) for metabolic syndrome and its components according to the quartiles of heme iron intake (mg/day) among men in the SDHS 2012–2013.

	Quartiles of Dietary Iron Intake, ORs (95% CI)				
	Q1	Q2	Q3	Q4	p-Value for Trend [1]
Heme Iron Intake (mg/day)	<0.93	(0.93–1.40)	(1.40–2.02)	≥2.02	
n	357	358	357	358	
Metabolic syndrome					
Crude model	Reference	1.27(0.89,1.81)	1.09(0.76,1.56)	1.21(0.85,1.73)	0.547
Adjusted model [2]	Reference	1.38(0.94,2.02)	1.12(0.75,1.68)	1.21(0.80,1.82)	0.415
Metabolic syndrome components					
Elevated blood pressure					
Crude model	Reference	0.99(0.76,1.28)	0.73(0.57,0.95) *	0.85(0.66,1.09)	0.059
Adjusted model [2]	Reference	1.29(0.95,1.75)	1.05(0.77,1.43)	1.23(0.89,1.69)	0.308
Elevated waist circumference					
Crude model	Reference	1.25(0.91,1.69)	1.37(1.01,1.86) *	1.82(1.36,2.45) **	0.001
Adjusted model [2]	Reference	1.26(0.91,1.76)	1.33(0.95,1.86)	1.89(1.35,2.65) **	0.003
Elevated fasting glucose					
Crude model	Reference	1.02(0.76,1.36)	0.85(0.63,1.14)	0.92(0.69,1.23)	0.597
Adjusted model [2]	Reference	1.12(0.82,1.54)	1.02(0.73,1.42)	1.18(0.84,1.65)	0.729
Elevated triglycerides					
Crude model	Reference	1.02(0.78,1.35)	1.06(0.81,1.39)	1.01(0.77,1.33)	0.978
Adjusted model [2]	Reference	0.90(0.67,1.22)	0.96(0.71,1.30)	0.82(0.59,1.12)	0.607
Reduced HDL-C					
Crude model	Reference	1.06(0.71,1.59)	1.03(0.69,1.54)	1.06(0.71,1.58)	0.991
Adjusted model [2]	Reference	1.05(0.68,1.61)	0.96(0.62,1.49)	0.98(0.62,1.55)	0.982

* $p < 0.05$, ** $p < 0.001$. [1] The p-value for the trend was examined using the medians in each quartile of dietary iron intake. [2] The models were adjusted for age, income, physical activity level, intentional physical exercise, smoking status, alcohol use, and dietary total energy intake.

Table 7. ORs (95% CI) for metabolic syndrome and its components according to the quartiles of nonheme iron intake (mg/day) among men in the SDHS 2012–2013.

	Quartiles of Dietary Iron Intake, ORs (95% CI)				
	Q1	Q2	Q3	Q4	*p*-Value for Trend [1]
Nonheme Iron Intake (mg/day)	<12.76	(12.76–16.31)	(16.31–21.72)	≥21.72	
n	358	357	357	358	
Metabolic syndrome					
Crude model	Reference	1.29(0.89,1.87)	1.65(1.15,2.37) **	1.60(1.11,2.30) *	0.026
Adjusted model [2]	Reference	1.49(0.99,2.25)	1.79(1.17,2.74) **	2.05(1.26,3.35) **	0.025
Metabolic syndrome components					
Elevated blood pressure					
Crude model	Reference	0.97(0.75,1.25)	1.16(0.91,1.50)	1.11(0.86,1.43)	0.443
Adjusted model [2]	Reference	1.21(0.89,1.64)	1.24(0.91,1.70)	1.28(0.88,1.84)	0.494
Elevated waist circumference					
Crude model	Reference	1.43(1.06,1.94) *	1.49(1.10,2.01) **	1.71(1.27,2.31) **	0.004
Adjusted model [2]	Reference	1.40(1.01,1.94) *	1.37(0.98,1.92)	1.48(1.01,2.17) *	0.143
Elevated fasting glucose					
Crude model	Reference	0.86(0.64,1.15)	0.94(0.71,1.25)	0.97(0.72,1.29)	0.769
Adjusted model [2]	Reference	1.02(0.74,1.42)	1.10(0.78,1.55)	1.33(0.89,1.98)	0.500
Elevated triglycerides					
Crude model	Reference	1.28(0.97,1.69)	1.37(1.04,1.79) *	1.31(0.99,1.73)	0.113
Adjusted model [2]	Reference	1.25(0.92,1.69)	1.23(0.90,1.69)	1.33(0.92,1.92)	0.399
Reduced HDL-C					
Crude model	Reference	1.04(0.70,1.57)	1.20(0.81,1.78)	1.06(0.71,1.58)	0.822
Adjusted model [2]	Reference	1.06(0.68,1.65)	1.38(0.87,2.20)	1.35(0.78,2.33)	0.495

* $p < 0.05$, ** $p < 0.001$. [1] The *p*-value for the trend was examined using the medians in each quartile of dietary iron intake. [2] The models were adjusted for age, income, physical activity level, intentional physical exercise, smoking status, alcohol use, and dietary total energy intake.

Table 8. ORs (95% CI) for metabolic syndrome and its components according to the quartiles of total iron intake (mg/day) among women in the SDHS 2012–2013.

	Quartiles of Dietary Iron Intake, ORs (95% CI)				
	Q1	Q2	Q3	Q4	*p*-Value for Trend [1]
Total Iron Intake (mg/day)	<11.67	(11.67–15.10)	(15.10–20.00)	≥20.00	
n	416	419	416	418	
Metabolic syndrome					
Crude model	Reference	1.02(0.74,1.40)	1.21(0.88,1.65)	0.96(0.70,1.32)	0.476
Adjusted model [2]	Reference	1.25(0.88,1.78)	1.41(0.98,2.02)	1.27(0.82,1.96)	0.322
Metabolic syndrome components					
Elevated blood pressure					
Crude model	Reference	0.86(0.68,1.10)	1.08(0.85,1.37)	0.85(0.67,1.09)	0.163
Adjusted model [2]	Reference	1.01(0.75,1.37)	1.29(0.94,1.76)	1.05(0.73,1.52)	0.306
Elevated waist circumference					
Crude model	Reference	0.97(0.76,1.24)	1.01(0.79,1.29)	0.93(0.72,1.19)	0.914
Adjusted model [2]	Reference	1.09(0.83,1.44)	1.13(0.85,1.50)	1.15(0.82,1.62)	0.839
Elevated fasting glucose					
Crude model	Reference	0.90(0.67,1.22)	1.23(0.92,1.64)	1.09(0.81,1.47)	0.216
Adjusted model [2]	Reference	1.01(0.73,1.41)	1.33(0.95,1.85)	1.23(0.83,1.83)	0.274
Elevated triglycerides					
Crude model	Reference	1.01(0.75,1.34)	1.13(0.85,1.50)	1.08(0.81,1.44)	0.814
Adjusted model [2]	Reference	1.06(0.77,1.44)	1.16(0.85,1.60)	1.18(0.81,1.72)	0.773
Reduced HDL-C					
Crude model	Reference	1.19(0.88,1.62)	1.14(0.84,1.56)	1.14(0.84,1.56)	0.710
Adjusted model [2]	Reference	1.37(0.98,1.91)	1.31(0.93,1.85)	1.40(0.93,2.10)	0.264

[1] The *p*-value for the trend was examined using medians in each quartile of dietary iron intake. [2] The models were adjusted for age, income, physical activity level, intentional physical exercise, smoking status, alcohol use, and dietary total energy intake.

Table 9. ORs (95% CI) for metabolic syndrome and its components according to the quartiles of heme iron intake (mg/day) among women in the SDHS 2012–2013.

	Quartiles of Dietary Iron Intake, ORs (95% CI)				
	Q1	Q2	Q3	Q4	p-Value for Trend [1]
Heme Iron Intake (mg/day)	<0.76	(0.76–1.19)	(1.19–1.81)	≥1.81	
n	416	419	419	415	
Metabolic syndrome					
Crude model	Reference	0.88(0.65,1.19)	0.81(0.60,1.11)	0.74(0.54,1.01)	0.273
Adjusted model [2]	Reference	1.08(0.78,1.52)	1.06(0.75,1.51)	1.08(0.74,1.57)	0.967
Metabolic syndrome components					
Elevated blood pressure					
Crude model	Reference	0.71(0.55,0.90) **	0.63(0.49,0.80) **	0.58(0.45,0.74) **	0.000
Adjusted model [2]	Reference	0.93(0.69,1.26)	0.90(0.67,1.23)	0.91(0.66,1.26)	0.924
Elevated waist circumference					
Crude model	Reference	0.94(0.74,1.21)	0.91(0.71,1.16)	0.77(0.59,0.99) *	0.193
Adjusted model [2]	Reference	1.18(0.90,1.54)	1.22(0.92,1.61)	1.05(0.78,1.42)	0.439
Elevated fasting glucose					
Crude model	Reference	0.85(0.63,1.13)	0.75(0.55,1.00)	0.87(0.65,1.17)	0.285
Adjusted model [2]	Reference	0.98(0.72,1.35)	0.94(0.67,1.30)	1.12(0.79,1.58)	0.755
Elevated triglycerides					
Crude model	Reference	1.06(0.79,1.41)	1.01(0.76,1.35)	1.00(0.75,1.34)	0.979
Adjusted model [2]	Reference	1.17(0.87,1.59)	1.14(0.83,1.57)	1.20(0.86,1.67)	0.700
Reduced HDL-C					
Crude model	Reference	1.13(0.83,1.55)	1.54(1.13,2.09) **	1.11(0.81,1.53)	0.033
Adjusted model [2]	Reference	1.24(0.89,1.73)	1.65(1.18,2.31) **	1.25(0.87,1.79)	0.027

* $p < 0.05$, ** $p < 0.001$. [1] The p-value for the trend was examined using medians in each quartile of dietary iron intake. [2] The models were adjusted for age, income, physical activity level, intentional physical exercise, smoking status, alcohol use, and dietary total energy intake.

Table 10. ORs (95% CI) for metabolic syndrome and its components according to the quartiles of nonheme iron intake (mg/day) among women in the SDHS 2012–2013.

	Quartiles of Dietary Iron Intake, ORs (95% CI)				
	Q1	Q2	Q3	Q4	*p*-Value for Trend [1]
Nonheme Iron Intake (mg/day)	<10.69	(10.69–13.76)	(13.76–18.24)	≥18.24	
n	418	418	416	417	
Metabolic syndrome					
Crude model	Reference	0.87(0.64,1.21)	1.25(0.92,1.70)	0.92(0.67,1.27)	0.111
Adjusted model [2]	Reference	1.01(0.71,1.44)	1.42(0.99,2.03)	1.14(0.74,1.75)	0.150
Metabolic syndrome components					
Elevated blood pressure					
Crude model	Reference	0.93(0.73,1.18)	1.17(0.92,1.49)	0.93(0.73,1.19)	0.214
Adjusted model [2]	Reference	1.10(0.81,1.49)	1.42(1.03,1.94) *	1.15(0.80,1.66)	0.140
Elevated waist circumference					
Crude model	Reference	0.90(0.70,1.15)	1.00(0.78,1.28) **	0.90(0.70,1.15)	0.689
Adjusted model [2]	Reference	0.99(0.75,1.30)	1.08(0.81,1.44)	1.05(0.74,1.47)	0.918
Elevated fasting glucose					
Crude model	Reference	0.96(0.71,1.31)	1.29(0.97,1.73)	1.16(0.86,1.56)	0.181
Adjusted model [2]	Reference	1.09(0.78,1.51)	1.40(1.00,1.95) *	1.29(0.87,1.91)	0.215
Elevated triglycerides					
Crude model	Reference	0.92(0.68,1.23)	1.24(0.94,1.64)	1.02(0.76,1.35)	0.194
Adjusted model [2]	Reference	0.94(0.68,1.28)	1.26(0.92,1.72)	1.06(0.73,1.54)	0.243
Reduced HDL-C					
Crude model	Reference	1.19(0.87,1.62)	1.16(0.85,1.58)	1.07(0.78,1.47)	0.685
Adjusted model [2]	Reference	1.30(0.93,1.81)	1.28(0.91,1.80)	1.23(0.82,1.85)	0.421

* $p < 0.05$, ** $p < 0.001$. [1] The *p*-value for the trend was examined using medians in each quartile of dietary iron intake. [2] The models were adjusted for age, income, physical activity level, intentional physical exercise, smoking status, alcohol use, and dietary total energy intake.

After stratifying by age group, a higher risk of MetS was observed in the third and fourth quartile of total dietary iron intake among the age group of above 60 (Table A1).

4. Discussion

The current findings were consistent with two previous studies indicating that the consumption of a certain type of dietary iron is related to the risk of developing MetS or its components [18,19]. Compared with the total dietary iron intake of the Western population, the average intake of Chinese adults is reportedly much higher, which is consistent with the iron intake determined in this study [18,20–22]. Free iron has strong pro-oxidant properties and generates reactive oxygen species by participating in Fenton chemistry, which results in the induction of oxidative damage and apoptosis [23]. Although iron plays an indispensable role in some physiological processes, excess iron can cause tissue damage or subclinical inflammation [24]. These side effects of iron are balanced out through binding to ferritin, an iron storage protein [13]. Given that several population-based human observational studies have reported a positive association between dietary iron intake and serum ferritin [10,11,21,25,26], the elevated concentration of serum ferritin is regarded as an indicator of iron overload. This research potentially supported the hypothesis that excessive iron in the body might play a causal role in the development of MetS. Nevertheless, MetS has been found to increase the inflammatory status and thereby induces changes in iron homeostasis; thus, the influence of MetS on iron metabolism cannot be ruled out [27,28]. These two mechanisms might mutually correspond to a continuous circle between iron metabolism and MetS.

In present study population, we observed that nonheme iron intake but not heme iron intake was associated with MetS. This finding was inconsistent with studies on Western populations, which concluded that heme iron intake is correlated with MetS [18,20,29]. Dietary iron is present in two forms: heme iron (animal food) and nonheme iron (plant and animal food) [13]. The Chinese diet is well known to be a predominantly plant-based diet. The major food sources of dietary iron in the current analysis were grains/potatoes, vegetables, and condiments. These foods are quite different from those in the typical diet of the Western population, in which a greater proportion of dietary iron is obtained from animal food [22], which indicated more heme iron intake. The dominance of plant-based food sources of dietary iron in the Chinese diet indicates a larger proportion of nonheme iron. In a Chinese study, no definite association was found for heme iron intake with iron status [25]. Another study on a Japanese population also found that heme iron intake is not correlated with iron status [30]. The difference between Western and Chinese dietary patterns could partially explain the difference in the associations of heme iron with MetS between the present findings and results from Western populations. Heme iron has been widely acknowledged to have better bioavailability than nonheme iron [31]. However, the ascorbic acid found in fruits and vegetables can enhance nonheme iron absorption [13]. In a study focusing on Chinese females, nonheme iron intake was found to be associated with serum ferritin [25]. This finding could support the correlation between nonheme iron intake and MetS observed in the present study.

In this study, higher risks of the four components of MetS, including elevated waist circumference, elevated triglycerides, reduced HDL-C, and elevated blood pressure, were observed with moderately higher total iron intake (the third quartile of total iron intake). Moderately higher nonheme iron intake (the third quartile of nonheme iron intake) was associated with higher risks of all five components of MetS. However, no associations were found between heme iron intake and the five components of MetS. Meanwhile, previous published studies indicated that iron overload in the body is correlated with all features of abnormal metabolism, such as obesity, hypertension, hyperlipidemia, and hyperglycemia [5,9,19,32,33].

We found that total iron and nonheme iron intake was associated with risk of MetS in male participants, but an association between iron intake and MetS was not observed in female participants in the current study. In the male population, the highest quartile of total iron intake was associated with a 2.1-fold higher risk of MetS compared with the lowest quartile. The Chinese recommended daily iron intakes for male and female adults are 12 mg and 20 mg, respectively, and the tolerable upper iron intake is 42 mg [34]. In this study, although the average dietary iron intake of both genders was far from the tolerable upper limit, the average total dietary iron intake of the male participants was almost double the recommended level, whereas that of the females was slightly under the recommended level. In addition, women experience greater iron loss than men due to menses. Increased iron stores are correlated with markers of chronic inflammation and risk of MetS [35,36]. The relatively higher iron intake and lower iron loss in men might explain the gender difference in the association between dietary iron intake and the occurrence of MetS observed in this study. Assessing the blood loss in the case of women warrants further study.

A limitation of this study is the methodology used to assess dietary intake. We used 3-day, 24-h dietary records to obtain food consumption information, and this information was then used to calculate the total, heme, and nonheme iron intake based on Chinese food composition data. Thus, the estimates of dietary iron intake were limited by the accuracy of the participants' recall and estimation. Furthermore, although we adjusted for several potential confounding factors, including dietary energy intake, we did not treat other iron intake-related dietary components, such as saturated fat or zinc, as covariates to avoid multicollinearity. Therefore, we cannot avoid the possibility that recall bias, nonresponse bias, and other unknown confounding factors might influence the result of the risk factor analysis. Finally, it is logical to hypothesize that dietary iron intake influences the body's metabolism, but the cross-sectional nature of the current study did not allow us to draw causal associations between total, heme, and nonheme dietary iron intake and MetS. Therefore, prospective observational studies or random clinical trials are needed to clarify the causal relationship between dietary iron and MetS.

5. Conclusions

Total and nonheme dietary iron intake was positively associated with MetS and its components in the adult population in metropolitan China, but an association between heme iron intake and MetS was not observed in this population. Moreover, a gender difference in the association between dietary iron intake and the occurrence of MetS was found: higher levels of total and nonheme iron intake were associated with greater risks of MetS in men but not in women.

Author Contributions: Conceptualization, G.D. and Z.Z.; Methodology, F.W. and Y.L.; Investigation, C.W., Z.W., C.G., X.J., J.Y. and H.P.; Data Analysis, J.Z. and Z.Z.; Writing-Original Draft Preparation, Z.Z.; Writing-Review & Editing, Y.H., J.S. and J.H.; Supervision and manuscript review, G.D. and F.W.

Funding: This study was supported by the National Key Research and Development Program of China (2016YFD0400602), the Three-year Action Plan on Public Health (15GWZK0801), Foundation of Shanghai Municipal Commission of Health and Family Planning (201740073) and DANONE International Prize for Nutrition (DIC2015-01).

Acknowledgments: We are grateful to all the individuals who participated in the study. We also thank our colleagues at the local districts' Centers for Disease Control and Prevention and local community health centers for their assistance with data collection.

Conflicts of Interest: The authors declare no conflict of interest.

Appendix A

Table A1. ORs (95% CI) for metabolic syndrome according to the quartiles of total iron intake (mg/day) by age group in the SDHS 2012–2013.

	Quartiles of Dietary Iron Intake, ORs (95% CI)				
	Q1	Q2	Q3	Q4	*p*-Value for Trend [1]
18–44 years					
Total Iron Intake (mg/day)	<12.84	(12.84–16.02)	(16.02–21.50)	≥21.50	
n	241	241	241	241	
Metabolic syndrome					
Crude model	Reference	1.55(0.81,2.94)	1.72(0.92,3.24)	1.88(1.00,3.51) *	0.237
Adjusted model [2]	Reference	1.53(0.77,3.03)	1.66(0.82,3.34)	1.96(0.86,4.46)	0.430
45–59 years					
Total Iron Intake (mg/day)	<13.03	(13.03–17.04)	(17.04–20.00)	≥22.31	
n	280	279	279	279	
Metabolic syndrome					
Crude model	Reference	1.10(0.74,1.65)	1.35(0.91,2.00)	1.05(0.70,1.58)	0.448
Adjusted model [2]	Reference	1.16(0.77,1.76)	1.41(0.92,2.16)	1.18(0.71,1.96)	0.458
60-years					
Total Iron Intake (mg/day)	<12.31	(12.31–16.28)	(16.28–20.00)	≥21.86	
n	255	254	255	254	
Metabolic syndrome					
Crude model	Reference	1.03(0.70,1.50)	1.11(0.77,1.62)	1.03(0.70,1.51)	0.949
Adjusted model [2]	Reference	1.24(0.83,1.86)	1.54(1.00,2.37) *	1.77(1.06,2.96) *	0.135

* $p < 0.05$. [1] The *p*-value for the trend was examined using the medians in each quartile of dietary iron intake. [2] The models were adjusted for sex, income, physical activity level, intentional physical exercise, smoking status, alcohol use, and dietary total energy intake.

References

1. Zhou, M.; Wang, H.; Zhu, J.; Chen, W.; Wang, L.; Liu, S.; Li, Y.; Wang, L.; Liu, Y.; Yin, P.; et al. Cause-specific mortality for 240 causes in China during 1990–2013: A systematic subnational analysis for the Global Burden of Disease Study 2013. *Lancet* **2016**, *387*, 251–272. [CrossRef]
2. Grundy, S.M.; Cleeman, J.I.; Daniels, S.R.; Donato, K.A.; Eckel, R.H.; Franklin, B.A.; Gordon, D.J.; Krauss, R.M.; Savage, P.J.; Smith, S.J.; et al. Diagnosis and management of the metabolic syndrome: An American Heart Association/National Heart, Lung, and Blood Institute Scientific Statement. *Circulation* **2005**, *112*, 2735–2752. [CrossRef] [PubMed]
3. Gu, D.; Reynolds, K.; Wu, X.; Chen, J.; Duan, X.; Reynolds, R.F.; Whelton, P.K.; He, J. Prevalence of the metabolic syndrome and overweight among adults in China. *Lancet* **2005**, *365*, 1398–1405. [CrossRef]
4. Li, Y.; Zhao, L.; Yu, D.; Wang, Z.; Ding, G. Metabolic syndrome prevalence and its risk factors among adults in China: A nationally representative cross-sectional study. *PLoS ONE* **2018**, *13*, e199293. [CrossRef] [PubMed]
5. Felipe, A.; Guadalupe, E.; Druso, P.; Carlos, M.; Pablo, S.; Oscar, C.; Luis, V.; Diego, M.; Jaime, R.; Inés, U.; Federico, L. Serum Ferritin Is Associated with Metabolic Syndrome and Red Meat Consumption. *Oxid. Med. Cell. Longev.* **2015**, *2015*, 1–8. [CrossRef] [PubMed]
6. Cho, M.; Park, J.; Choi, W.; Cho, A.; Lee, Y. Serum ferritin level is positively associated with insulin resistance and metabolic syndrome in postmenopausal women: A nationwide population-based study. *Maturitas* **2017**, *103*, 3–7. [CrossRef] [PubMed]
7. Kilani, N.; Vollenweider, P.; Waeber, G.; Marques-Vidal, P. Iron metabolism and incidence of metabolic syndrome. *Nutr. Metab. Cardiovasc. Dis.* **2015**, *25*, 1025–1032. [CrossRef] [PubMed]
8. Jehn, M.; Clark, J.M.; Guallar, E. Serum ferritin and risk of the metabolic syndrome in U.S. adults. *Diabetes Care* **2004**, *27*, 2422–2428. [CrossRef] [PubMed]
9. Fleming, R.E.; Ponka, P. Iron overload in human disease. *N. Engl. J. Med.* **2012**, *366*, 348–359. [CrossRef] [PubMed]
10. Cade, J.E.; Moreton, J.A.; O'Hara, B.; Greenwood, D.C.; Moor, J.; Burley, V.J.; Kukalizch, K.; Bishop, D.T.; Worwood, M. Diet and genetic factors associated with iron status in middle-aged women. *Am. J. Clin. Nutr.* **2005**, *82*, 813–820. [CrossRef] [PubMed]
11. Fleming, D.J.; Jacques, P.F.; Dallal, G.E.; Tucker, K.L.; Wilson, P.W.; Wood, R.J. Dietary determinants of iron stores in a free-living elderly population: The Framingham Heart Study. *Am. J. Clin. Nutr.* **1998**, *67*, 722–733. [CrossRef] [PubMed]
12. Ferreira, C.; Santambrogio, P.; Martin, M.E.; Andrieu, V.; Feldmann, G.; Henin, D.; Beaumont, C. H ferritin knockout mice: A model of hyperferritinemia in the absence of iron overload. *Blood* **2001**, *98*, 525–532. [CrossRef] [PubMed]
13. Geissler, C.; Singh, M. Iron, Meat and Health. *Nutrients* **2011**, *3*, 283–316. [CrossRef] [PubMed]
14. Monsen, E.R.; Hallberg, L.; Layrisse, M.; Hegsted, D.M.; Cook, J.D.; Mertz, W.; Finch, C.A. Estimation of available dietary iron. *Am. J. Clin. Nutr.* **1978**, *31*, 134. [CrossRef] [PubMed]
15. Yang, Y.; Wang, G.; Pan, X. *Chinese Food Composition Book 2*; Peking University Medical Press: Beijing, China, 2005; pp. 75–216, ISBN 9787810716789.
16. Yang, Y.; Wang, G.; Pan, X. *Chinese Food Composition Book 1*; Peking University Medical Press: Beijing, China, 2009; pp. 45–192, ISBN 9787811167276.
17. Pretorius, B.; Schonfeldt, H.C.; Hall, N. Total and haem iron content lean meat cuts and the contribution to the diet. *Food Chem.* **2016**, *193*, 97–101. [CrossRef] [PubMed]
18. De Oliveira Otto, M.C.; Alonso, A.; Lee, D.; Delclos, G.L.; Bertoni, A.G.; Jiang, R.; Lima, J.A.; Symanski, E.; Jacobs, D.R.; Nettleton, J.A. Dietary Intakes of Zinc and Heme Iron from Red Meat, but Not from Other Sources, Are Associated with Greater Risk of Metabolic Syndrome and Cardiovascular Disease. *J. Nutr.* **2012**, *142*, 526–533. [CrossRef] [PubMed]
19. Choi, J.S.; Koh, I.; Lee, H.J.; Kim, W.H.; Song, J. Effects of excess dietary iron and fat on glucose and lipid metabolism. *J. Nutr. Biochem.* **2013**, *24*, 1634–1644. [CrossRef] [PubMed]
20. Dos Santos Vieira, D.; Hermes Sales, C.; Galvão Cesar, C.; Marchioni, D.; Fisberg, R. Influence of Haem, Non-Haem, and Total Iron Intake on Metabolic Syndrome and Its Components: A Population-Based Study. *Nutrients* **2018**, *10*, 314. [CrossRef] [PubMed]

21. Young, I.; Parker, H.; Rangan, A.; Prvan, T.; Cook, R.; Donges, C.; Steinbeck, K.; O'Dwyer, N.; Cheng, H.; Franklin, J.; et al. Association between Haem and Non-Haem Iron Intake and Serum Ferritin in Healthy Young Women. *Nutrients* **2018**, *10*, 81. [CrossRef] [PubMed]
22. Samaniego-Vaesken, M.; Partearroyo, T.; Olza, J.; Aranceta-Bartrina, J.; Gil, Á.; González-Gross, M.; Ortega, R.; Serra-Majem, L.; Varela-Moreiras, G. Iron Intake and Dietary Sources in the Spanish Population: Findings from the ANIBES Study. *Nutrients* **2017**, *9*, 203. [CrossRef] [PubMed]
23. Hansen, J.B.; Moen, I.W.; Mandrup-Poulsen, T. Iron: The hard player in diabetes pathophysiology. *Acta Physiol.* **2014**, *210*, 717–732. [CrossRef] [PubMed]
24. Day, S.M.; Duquaine, D.; Mundada, L.V.; Menon, R.G.; Khan, B.V.; Rajagopalan, S.; Fay, W.P. Chronic iron administration increases vascular oxidative stress and accelerates arterial thrombosis. *Circulation* **2003**, *107*, 2601–2606. [CrossRef] [PubMed]
25. He, J.; Shen, X.; Fang, A.; Song, J.; Li, H.; Guo, M.; Li, K. Association between predominantly plant-based diets and iron status in Chinese adults: A cross-sectional analysis. *Br. J. Nutr.* **2016**, *116*, 1621–1632. [CrossRef] [PubMed]
26. Davis, C.D.; Malecki, E.A.; Greger, J.L. Interactions among dietary manganese, heme iron, and nonheme iron in women. *Am. J. Clin. Nutr.* **1992**, *56*, 926–932. [CrossRef] [PubMed]
27. MacDonald, M.J.; Cook, J.D.; Epstein, M.L.; Flowers, C.H. Large amount of (apo)ferritin in the pancreatic insulin cell and its stimulation by glucose. *FASEB J.* **1994**, *8*, 777–781. [CrossRef] [PubMed]
28. Deugnier, Y.; Bardou-Jacquet, E.; Le Lan, C.; Brissot, P. Hyperferritinemia not related to hemochromatosis. *Gastroenterol. Clin. Biol.* **2009**, *33*, 323. [CrossRef] [PubMed]
29. Azadbakht, L.; Esmaillzadeh, A. Red meat intake is associated with metabolic syndrome and the plasma C-reactive protein concentration in women. *J. Nutr.* **2009**, *139*, 335–339. [CrossRef] [PubMed]
30. Asakura, K.; Sasaki, S.; Murakami, K.; Takahashi, Y.; Uenishi, K.; Yamakawa, M.; Nishiwaki, Y.; Kikuchi, Y.; Takebayashi, T.; Group, J.D.S.S.; et al. Iron intake does not significantly correlate with iron deficiency among young Japanese women: A cross-sectional study. *Public Health Nutr.* **2009**, *12*, 1373–1383. [CrossRef] [PubMed]
31. Hallberg, L. Bioavailability of dietary iron in man. *Annu. Rev. Nutr.* **1981**, *1*, 123–147. [CrossRef] [PubMed]
32. Piperno, A.; Trombini, P.; Gelosa, M.; Mauri, V.; Pecci, V.; Vergani, A.; Salvioni, A.; Mariani, R.; Mancia, G. Increased serum ferritin is common in men with essential hypertension. *J. Hypertens.* **2002**, *20*, 1513–1518. [CrossRef] [PubMed]
33. Bozzini, C.; Girelli, D.; Olivieri, O.; Martinelli, N.; Bassi, A.; De Matteis, G.; Tenuti, I.; Lotto, V.; Friso, S.; Pizzolo, F.; et al. Prevalence of body iron excess in the metabolic syndrome. *Diabetes Care* **2005**, *28*, 2061–2063. [CrossRef] [PubMed]
34. Chinese Nutrition Society. *Chinese Dietary Reference Intakes*; Science Press: Beijing, China, 2013.
35. Vari, I.S.; Balkau, B.; Kettaneh, A.; André, P.; Tichet, J.; Fumeron, F.; Caces, E.; Marre, M.; Grandchamp, B.; Ducimetière, P. Ferritin and Transferrin Are Associated with Metabolic Syndrome Abnormalities and Their Change Over Time in a General Population. *Diabetes Care* **2007**, *30*, 1795. [CrossRef] [PubMed]
36. Fumeron, F.; Pean, F.; Driss, F.; Balkau, B.; Tichet, J.; Marre, M.; Grandchamp, B. Ferritin and transferrin are both predictive of the onset of hyperglycemia in men and women over 3 years: The data from an epidemiological study on the Insulin Resistance Syndrome (DESIR) study. *Diabetes Care* **2006**, *29*, 2090–2094. [CrossRef] [PubMed]

nutrients

MDPI

Article

Iron Deficiency and Neuroendocrine Regulators of Basal Metabolism, Body Composition and Energy Expenditure in Rats

Jorge Moreno-Fernandez, Javier Díaz-Castro *, María J. M. Alférez and Inmaculada López-Aliaga

Department of Physiology (Faculty of Pharmacy, Campus Universitario de Cartuja) and Institute of Nutrition and Food Technology "José Mataix", University of Granada, E-18071 Granada, Spain; jorgemf@ugr.es (J.M.-F.); malferez@ugr.es (M.J.M.A.); milopez@ugr.es (I.L.-A.)

* Correspondence: javierdc@ugr.es; Tel.: +34-958-241-000 (ext. 20303)

Received: 11 February 2019; Accepted: 12 March 2019; Published: 15 March 2019

Abstract: Although dietary iron is a determinant of iron status in animals, body fat mass has been reported to have an inverse association with iron status in human studies. The goal of this study was to determine the relationship between Fe homeostasis, body composition, energy expenditure and neuroendocrine regulators for severe Fe-deficiency anaemia. Forty male Wistar albino rats recently weaned were divided at random into two groups: the control group was fed the basal diet, AIN-93G diet (normal-Fe) and the anaemic group received a low-Fe diet for 40 days. Neuroendocrine parameters that regulate basal metabolism and appetite (thyroid hormones, ghrelin, glucose-dependent insulinotropic polypeptide (GIP), glucagon, insulin, adrenocorticotropic hormone and corticosterone), body composition, respiratory volumes, energy expenditure, haematological and biochemical were assessed. Total body fat was lower, whereas lean mass, free and total water were higher in the anemic group. O_2 consumption, CO_2 production, energy expenditure (EE) and respiratory quotient (RQ) were lower in the Fe-deficient animals. Triiodothyronine and thyroxine hormones decreased, while thyroid-stimulating hormone increased in the anemic group. Circulating levels of ghrelin were lower in the anemic group, while GIP, glucagon, insulin, corticosterone and adrenocorticotropic hormone levels were higher. Fe-deficiency impairs weight gain in the rats, with marked reductions in lean mass and body fat, indicating lower energy stores.

Keywords: ferropenic anaemia; endocrine regulators; body composition; energy expenditure

---

1. Introduction

Iron-deficiency Anaemia (IDA) occurs when Fe loss and body's requirement for iron is not met by dietary sources such that Fe storage of the organism is depleted. This pathological process is characterized by the production of smaller red cells because the concentration of hemoglobin (Hb) is abnormally low [1].

While dietary Fe is a determinant of Fe status, an inverse association between fat mass and Fe status has been previously reported [2]. While studies have shown that increased fat mass might increase the risk of IDA, this is due to the fact that ferritin is an acute-phase protein that is elevated by the low-grade inflammation that occurs when adipose tissues are enlarged [3]. This fact is due that ferritin is an acute-phase protein that may be elevated by the low-grade inflammation that occurs when adipose tissues are enlarged [3]. However, there is scarce information about body composition changes in situation of severe IDA. Given that the burden of chronic diseases is rapidly increasing in developing countries as well as the existence of factors such as deficient diets [4], which influence Fe status and adiposity, it is important to have a good understanding of the relationship between Fe homeostasis and body composition. On the other hand, as the liver is the main Fe storage organ in the organism and it is vital in the regulation of iron homeostasis, the progression of iron depletion during iron deficiency is of great interest.

Metabolic rate is mainly regulated by the central nervous system, which senses energy balance from a wide range of humoral and neural signals, and controls energy intake and basal metabolism [5]. Ghrelin and glucose-dependent insulinotropic polypeptide are two hormones produced mainly by the gut and are secreted from the gastrointestinal tract in response to a meal. Gut hormone concentrations change with weight loss and nutritional deficiencies [6]. Scientific reports have shown that iron deficiency is positively associated with adiposity [7]. In this sense, a decreased concentration of iron, and consecutively haemoglobin, ferritin and decreased level of saturated transferrin, were observed in obese individuals more often than in normal weight subjects [8]. Loss of appetite, retarded growth and development are typical manifestations of the nutritional ferropenic anemia; however, hormonal control for body composition changes during IDA has been not yet fully explained.

Therefore, the current study elucidated the interactions between basal metabolism hormones (ghrelin, GIP, insulin, cortisol, thyroid hormones) haematological status, body composition, and energy expenditure during severe iron deficiency anaemia in young rats.

2. Material and Methods

2.1. Study Animals

Forty male Wistar albino rats recently weaned at 21 days of life and average weight of 42 g were used. Animals were housing at the Laboratory Animals Unit of the Center of Biomedical Research of the University of Granada, certified as free of pathogens and the animals were housed at conditions of high biological safety, with sanitary and environmental rigorously controlled parameters. The Unit's housing and handling conditions were approved by the Ethics Committee of the University of Granada (Ref. 11022011) and abide with the Recommendations of the European Community guidelines (Declaration of Helsinki; Directive 2010/63/EU for animal experiments).

2.2. Study Design and Diet

After weaning, rats were placed on an experimental period (EP) of 40 days and randomly divided into two groups, a control group fed AIN-93G diet [9] and an anaemic group, receiving the same diet, but with a low-Fe content (5 mg Fe/kg diet) [10]. Nutrient composition of the experimental diet is shown in Table 1. The analysis of content of Fe in the diet (mg/kg) was as follows: 44.82 for normal-Fe diet and 6.21 for low-Fe diet. Diet intake was controlled, pair feeding all the animals (80% of the average intake) and deionized water was available ad libitum. Animals were housed in individual metabolic cages in an environmentally controlled room on a 12 light/12 dark cycle (9:00 a.m.–9:00 p.m.), at 22 ± 1 °C and humidity of 55–65%. Feed and mineral-free water were provided ad libitum. At the end of EP, whole body composition and respiratory volumes and flows and energy expenditure were assessed. At the end of the EP, whole body composition and respiratory volumes and flows and energy expenditure were assessed as described in 2.10–2.12. After fasting overnight, animals were weighed and anesthetized by intraperitoneal injection of sodium pentobarbital (5 mg/100 g body weight) (Sigma, St Louis, MO, USA). After midline laparotomy, the rats were bled from the abdominal aorta, one aliquot of blood was collected in tubes with EDTA anticoagulant for haematological analysis. Serum was separated from the EDTA-free aliquot by centrifugation at 1500× g for 10 min at a 4 °C and stored at −80 °C to measure ferritin, transferrin saturation, total iron binding capacity (TIBC) and biochemical parameters. Liver was removed, rinsed in ice-cold normal saline solution (0.9%, w/v, NaCl), immediately weighed and stored at −20 °C for Fe analysis.

Table 1. Composition of the experimental diets.

Component	g/Kg Diet (Dry Weight)
Protein (casein)	210
Fat (olive oil)	100
Fibre (micronized cellulose)	50
Mineral supplement [a]	35
Vitamin supplement [b]	10
Choline chloride	2
Wheat starch	491
Sucrose	100
L-cystine	2
Energy (kJ/kg)	17940

[a] During the experimental period (EP), the mineral premix was prepared according to the recommendations of the AIN-93 G Diet [9] for control rats (normal-Fe: 45 mg/kg diet) and for anaemic rats (low-Fe: 5 mg/kg diet) [10].
[b] The vitamin premix were prepared according to the recommendations of the AIN-93 G Diet [9] for growing rats.

2.3. Dry Matter

Determination of water content (in diet and liver) was carried out by drying the material to a constant weight at 105 + 2 °C in an oven (~48 h).

2.4. Iron Determination

The diet and liver samples were previously mineralized by wet method in a sand bath (J.R. Selecta, Barcelona, Spain); the samples were placed in a resistant flask and dissolved using nitric acid followed by a mixture of HNO_3:$HClO_4$ (69%:70%, *v*/*v*; Merck KGaA, Darmstadt, Germany; ratio 1:4, *v*/*v*) until the total elimination of organic matter. Finally, the samples were diluted with bidistilled ultrapure Milli-Q water. Fe analysis was undertaken using an Optima 8300 (PerkinElmer Inc. Waltham, MA, USA) inductively coupled plasma-optical emission spectrometer (ICP-OES) with a segmented-array charge-coupled Device (SCD) high-performance detector. Fe was analysed in liver and diets according to compatibility under optimised set of conditions. For the calibration of the apparatus, multi elemental Astasol calibration solutions (Analytika, Khodlova, Prague) were used. For the calibration curve, the following working dilutions of the analytical standard were prepared: 0.1, 0.5, 1.0, 10, 50 mg/L. An internal standard solution of 10 mg L^{-1} was used after each series of five samples. The acceptable result was assessed as 10%. The samples were measured in three replicates.

2.5. Haematological Test

Haemoglobin (Hb) concentration, red blood cells (RBCs), haematocrit, mean corpuscular volume (MCV), mean corpuscular Hb (MCH), mean corpuscular Hb concentration (MCHC), red cell distribution width (RDW), platelets, white blood cells (WBCs) and lymphocytes, of fresh blood samples were measured using an automated hematology analyzer Mythic 22CT (C2 Diagnostics, Grabels, France).

2.6. Serum Iron, Total Iron Binding Capacity and Transferrin Saturation

Serum iron and total iron binding capacity (TIBC) were determined using a quantitative colorimetric enzymatic assay (Sigma Diagnostics Iron and Total Iron- Binding Capacity reagents, Sigma Diagnostics, St Louis, MO, USA). The absorbance was read at 550 nm on a microplate reader (Bio-Rad Laboratories Inc., CA, USA). Transferrin saturation percentage was calculated by dividing serum iron by TIBC, and then multiplying by 100.

2.7. Serum Ferritin

Serum ferritin concentration was determined using the Rat Ferritin ELISA Kit (Biovendor Gmbh, Heidelberg, Germany). The absorbance of the reaction mixtures was read in a microplate plate reader at 450 nm (reference 650 nm) using a Bio-Rad microplate reader (Bio-Rad Laboratories Inc., CA, USA). The intensity of the color was inversely proportional to the serum ferritin concentration.

2.8. Serum Hepcidin

Hepcidin-25 concentration was determined using a DRG ELISA Kit (DRG Instruments GmbH, Marburg, Germany). The microtiter wells were coated with a monoclonal (mouse) antibody directed toward an antigenic site of the hepcidin-25 molecule. Endogenous hepcidin-25 of a sample competed with a hepcidin-25-biotin conjugate for binding to the coated antibody. After incubation, the unbound conjugate was washed off and a streptavidin-peroxidase enzyme complex was added to each well. Substrate was added to detect antigen–antibody–enzyme complex and development of blue color. The microplate was read at 450 nm on a microplate reader (Bio-Rad Laboratories Inc.) and the intensity of color developed was reverse proportional to the concentration of hepcidin in the sample. Results were expressed in nanograms per milliliter of serum.

2.9. Biochemical Parameters

Serum total protein, albumin, total cholesterol, LDL-cholesterol, triglycerides, glucose, transaminases (AST and ALT), bilirubin, urea, creatinine, amylase, cortisol and creatine kinase-MB were measured by standard colorimetric and enzymatic methods, using a BS-200 Chemistry Analyzer (Shenzhen Mindray Bio-Medical Electronics Co. Ltd., Shenzhen, China). Two replicate samples were analyzed for measure each parameter, and the averages in each paired data were determined.

2.10. Assessment of Body Composition

Whole body composition (fat and lean tissues) was determined using quantitative magnetic resonance (QMR) with an Echo MRI Analyzer system by Echo Medical Systems (Houston, TX, USA) [11]. All QMR measurements were made during the light phase (09:00 A.M.–6:00 P.M.). Scans were performed by placing animals into a thin-walled plastic cylinder (3 mm thick, 6.5 cm inner diameter), with a cylindrical plastic insert added to limit movement. While in the tube, animals were briefly subjected to a low-intensity (0.05 Tesla) electromagnetic field to measure fat, lean mass, free water, and total body water. Briefly, this system generates a signal that modifies the spin patterns of hydrogen atoms within the subject, and uses an algorithm to evaluate the four components measured—fat mass, lean muscle mass equivalent, total body water, and free water. QMR scans were performed with accumulation times of 2 min.

2.11. Respiratory Gas Collection, Energy Expenditure Calculation and Analysis

The instruments used for measurement of the respiratory quotient in the rats consisted of acrylic metabolic chambers, gas analyzers (model LE 405 Gas Analyzer Panlab Technology for Bioresearch, Madrid, Spain), and a switching system (model L 400 Air Supply and Switching, Panlab Technology for Bioresearch). Prior to data collection, the system was calibrated using certified gas cylinders (Gilmore Liquid Air Co., South El Monte, CA, USA) containing 50% O_2/5% CO_2/45% N_2 (high point calibration) and 20% O_2/0% CO_2/80% N_2 (low point calibration) per manufacturer's instructions. A clear, metabolic chamber, containing a small amount of bedding material, was placed on a over the Pyrex pan. The ambient chamber temperature was maintained at 22 $\pm$ 2 °C. All the weights were entered into the derived inlet flow rate equation to estimate the appropriate value. Individual rats were placed in the metabolic chamber and allowed to acclimate for 15 min prior to data collection. During this period, room air was allowed to flow through the chamber. After acclimation, respiratory gas analysis was carried out utilizing Metabolism H software (Panlab, S.L., Barcelona, Spain) running on a HP Compaq Pentium 4 computer (Hewlett-Packard Company, Palo Alto, CA, USA). The data collection paradigm consisted of (1) a 15-min acclimation period during which time room air was flowing through the metabolic chamber, (2) a 1-min room air reference sampling period, and (3) a 10-min chamber air sampling period during 24 h. This paradigm yielded enough time for the small volumes of gases exchanged by the rat to reach equilibrium within the chamber, accommodating the inherently slow sampling process associated with low inlet and sampling flow rates. The mean values of VO_2, VCO_2, respiratory quotient (RQ) and daily energy expenditure (EE) over the 10-min chamber sampling period were then calculated by the Metabolism H software (Harvard Apparatus, Massachusetts, MO, USA). The total time the rats spent on the system was 24 h each. Metabolic data were collected from individual rats of each experimental group and were analyzed during the day (tested between 9:00 a.m. and 9:00 p.m., and night (tested between 9:00 p.m. and 9:00 a.m.) to explore the possible influence of time (during the light phase) on the studied metabolic parameters. By utilizing an open-circuit system that pulled air into the chamber under slightly negative pressure to maintain a constant flow rate, the influence of the rat's physical activity on internal chamber pressure (and thus on the ability to measure true VO_2) was avoided.

2.12. Thyroid Hormones, Ghrelin, Glucose-Dependent Insulinotropic Polypeptide, Glucagon, Insulin, Adrenocorticotropic Hormone and Corticosterone Measurement

Triiodothyronine (T_3), thyroxine (T_4) and thyroid-stimulating hormone (TSH), were determined using the RTHYMAG-30K Milliplex MAP Rat Thyroid Magnetic Bead Panel; ghrelin (active), GIP (total), glucagon and insulin, were determined using the RMHMAG-84K Milliplex MAP Rat Metabolic Hormone Magnetic Bead Panel; adrenocorticotropic hormone (ACTH) and corticosterone plasma levels were determined using the RSHMAG-69K Milliplex MAP Rat Stress Hormone Magnetic Bead Panel (Millipore Corporation, MO, Massachusetts, USA), based on immunoassays on the surface of fluorescent-coded beads (microspheres), following the specifications of the manufacturer (50 events per bead, 50 µL sample, gate settings: 8000–15,000, time out 60 s, bead set: 34). Plate was read on LABScan 100 analyzer (Luminex Corporation, Austin, TX, USA) with xPONENT software (MO, Massachusetts, USA) for data acquisition. Average values for each set of duplicate samples or standards were within 15% of the mean. Thyroid hormones, ghrelin, GIP, glucagon, insulin ACTH and corticosterone concentrations in plasma samples were determined by comparing the mean of duplicate samples with the standard curve for each assay.

2.13. Statistical Analysis

Statistical analysis was performed with SPSS version 20.0 (SPSS Inc., Chicago, IL, USA) software package. Data were expressed as means ± standard error of the mean (SEM). Prior to perform any statistical analysis, all variables were checked to assess the equality of variances (homogeneity of variance) using the Levene's test. Student's *t*-test for independent samples was used to check the difference in mean between control vs. anaemic groups. A value of $p < 0.05$ was considered statistically significant.

3. Results

3.1. Haematological and Bichemical Parameters

After Fe deprivation, differences in were found in all the haematological parameters analysed on both experimental groups. Most of the haematological parameters were drastically low ($p < 0.001$): Hb concentration, RBCs, haematocrit, MCV, MCH, MCHC serum Fe, transferrin saturation, serum ferritin, and serum hepcidin. On the other hand, platelets, RDW and TIBC increased markedly ($p < 0.001$), while no changes were recorded in WBCs (Table 2). All of these results indicate that the Fe-deficiency anaemia was experimentally induced in the rats. With regard to the biochemical parameters, albumin, total cholesterol, triglycerides, glucose, AST, ALT, bilirubin, urea, amylase and cortisol increased markedly in the Fe-deficient group in comparison with the control group ($p < 0.001$) (Table 2).

Table 2. Haematological and biochemical parameters of control and anaemic rats.

Haematological Parameters	Normal-Fe Control Group (n = 20)	Low-Fe Anaemic Group (n = 20)
Hb concentration (g/L)	138.83 ± 3.19	60.36 ± 3.22 *
RBCs (10^{12}/L)	7.05 ± 0.19	3.02 ± 0.24 *
Haematocrit (%)	39.87 ± 1.14	11.61 ± 1.32 *
MCV (fL)	56.40 ± 0.55	38.26 ± 0.37 *
MCH (pg)	19.91 ± 0.14	14.32 ± 0.68 *
MCHC (g/dL)	35.24 ± 0.36	30.88 ± 0.87 *
RDW (%)	16.67 ± 0.34	19.31 ± 0.42 *
Platelets (10^9/L)	745 ± 73.19	2251 ± 118 *
WBCs (10^9/L)	8.88 ± 0.38	8.26 ± 0.97
Lymphocytes (10^6/mL)	8.01 ± 0.61	5.88 ± 0.85 *
Serum Fe (μg/L)	1350 ± 106	598 ± 57.01 *
TIBC (μg/L)	2756 ± 197	18235± 676 *
Transferrin saturation (%)	48.55 ± 6.49	3.75 ± 0.41 *
Serum ferritin (μg/L)	81.45 ± 2.37	49.12 ± 1.48 *
Serum hepcidin (ng/mL)	16.98 ± 0.45	12.51 ± 0.62 *
Biochemical parameters		
Total protein (g/dL)	4.97 ± 0.15	5.27 ± 0.13
Albumin (g/dL)	2.84 ± 0.04	3.34 ± 0.12 *
Total cholesterol (mg/dL)	88.92 ± 7.45	109.27 ± 9.69 *
LDL-cholesterol (U/L)	2815 ± 444	2829 ± 446
Triglycerides (mg/dL)	73.92 ± 3.37	217.52 ± 28.46 *
Glucose (mg/dL)	69.61 ± 3.97	86.13 ± 3.68 *
AST (IU/L)	103.58 ± 8.93	228.04 ± 18.45 *
ALT (IU/L)	24.57 ± 1.16	52.28 ± 2.73 *
Bilirubin (mg/dL)	0.81 ± 0.09	1.25 ± 0.13 *
Urea (mg/dL)	33.32 ± 1.77	42.47 ±2.01 *
Creatinine (mg/dL)	0.048 ± 0.017	0.043 ± 0.015
Amilase (U/L)	76.40 ± 5.16	379.35 ± 16.81 *
Cortisol (μg/L)	21.51± 1.07	42.21± 1.88 *
Creatine kinase-MB (U/L)	1531 ± 112	1391 ± 112

Mean values ± SEM. Hb haemoglobin, RBCs red blood cells, MCV mean corpuscular volume, MCH mean corpuscular Hb, MCHC mean corpuscular Hb concentration, RDW red cell distribution width, WBCs white blood cells, TIBC total Fe-binding capacity, LDL-cholesterol low-density lipoprotein cholesterol, AST aspartate aminotransferase, ALT alanine aminotransferase. * Significantly different ($p < 0.001$) from the control group by Student's t-test.

3.2. Hepatosomatic Index

Body weight was lower in anaemic animals as compared to the control animals ($p < 0.001$), and liver weight did not change significantly. As a consequence, the hepatosomatic index (ratio of liver weight to body weight) was higher in the Fe-deficient animals ($p < 0.05$). The differences in the hepatic Fe content on day 40 of the study were markedly pronounced ($p < 0.001$) and directly correlated with the Fe-restriction in the diet (Table 3).

Table 3. Hepatosomatic index and liver Fe content in control and anemic rats.

	Normal-Fe Control Group (n = 20)	Low-Fe Anaemic Group (n = 20)
Body weight (g)	242.7 ± 4.6	200.8 ± 2.6 **
Liver weight (g)	6.415 ± 0.26	6.032 ± 0.34
Liver weight/body weight (%)	2.59 ± 0.06	2.83 ± 0.09 *
Liver Fe content (μg/g dry weight)	609.26 ± 34.12	424.12 ± 23.10 **

Mean values ± SEM. * Significantly different ($p < 0.05$) from the control group by Student's t-test. ** Significantly different ($p < 0.001$) from the control group by Student's t-test.

3.3. Body Composition

Marked differences were found in body composition parameters between both experimental groups. Body fat was lower in the anaemic group compared with the control group ($p < 0.001$). In contrast, lean mass, free water and total water were higher in the Fe-deficient group ($p < 0.01$ for lean mass and $p < 0.001$ for free and total water) (Table 4).

Table 4. Body composition in control and anemic rats.

	Normal-Fe Control Group (n = 20)	**Low-Fe Anaemic Group (n = 20)**
Fat (%)	7.29 ± 0.36	6.38 ± 0.41 **
Fat (g) [1]	22.20 ± 2.35	15.62 ±1.57 **
Lean mass (%)	90.63 ± 0.47	92.51 ± 0.63 *
Lean mass (g) [1]	263.52 ± 4.80	242.17 ± 3.69 *
Free water (%)	0.42 ± 0.04	0.76 ± 0.09 *
Free water (g) [1]	1.24 ± 0.21	1.12 ± 0.21 *
Total water (%)	76.79 ± 0.57	79.89 ± 0.45 **
Total water (g) [1]	223.31 ± 4.09	206.47 ± 3.09 **

Mean values ± SEM. * Significantly different ($p < 0.01$) from the control group by Student's t-test. ** Significantly different ($p < 0.001$) from the control group by Student's t-test. [1] Includes water contained in the tissue.

3.4. Respiratory Volumes and Flows, Energy Expenditure, O_2 Consumption

With regard to respiratory volumes and flows, EE, O_2 consumption was lower in the Fe-deficient rats in both periods (day and night) ($p < 0.001$), CO_2 production was also lower in the anaemic group ($p < 0.001$ during the day and $p < 0.01$ during the night), EE decreased in the Fe-deficient group ($p < 0.001$ in both periods) and RQ also decreased ($p < 0.01$ in both periods) (Table 5, Figure 1).

Table 5. O_2 consumption, CO_2 production, energy expenditure and respiratory quotient in control and anaemic rats.

	Normal-Fe Control Group (n = 20)	**Low-Fe Anaemic Group (n = 20)**
Day period		
VO_2 (mL/min)	8.86 ± 0.15	7.43 ± 0.09 **
VCO_2 (mL/min)	7.91 ± 0.20	6.07± 0.32 **
EE (kcal/day)	62.74 ± 0.95	51.59 ± 0.97 **
RQ	0.89 ± 0.02	0.82 ± 0.03*
Night period		
VO_2 (mL/min)	8.31 ± 0.39	7.36 ± 0.26 **
VCO_2 (mL/min)	7.12 ± 0.18	53.15 ± 1.34 **
EE (kcal/day)	59.49 ± 1.39	6.45 ± 0.37 *
RQ	0.86 ± 0.01	0.81 ± 0.02 *

Mean values ± SEM. *VO_2* O_2 consumption, *VCO_2* CO_2 production, *EE* energy expenditure, *RQ* respiratory quotient; * Significantly different ($p < 0.01$) from the control group by Student's t-test. ** Significantly different ($p < 0.001$) from the control group by Student's t-test.

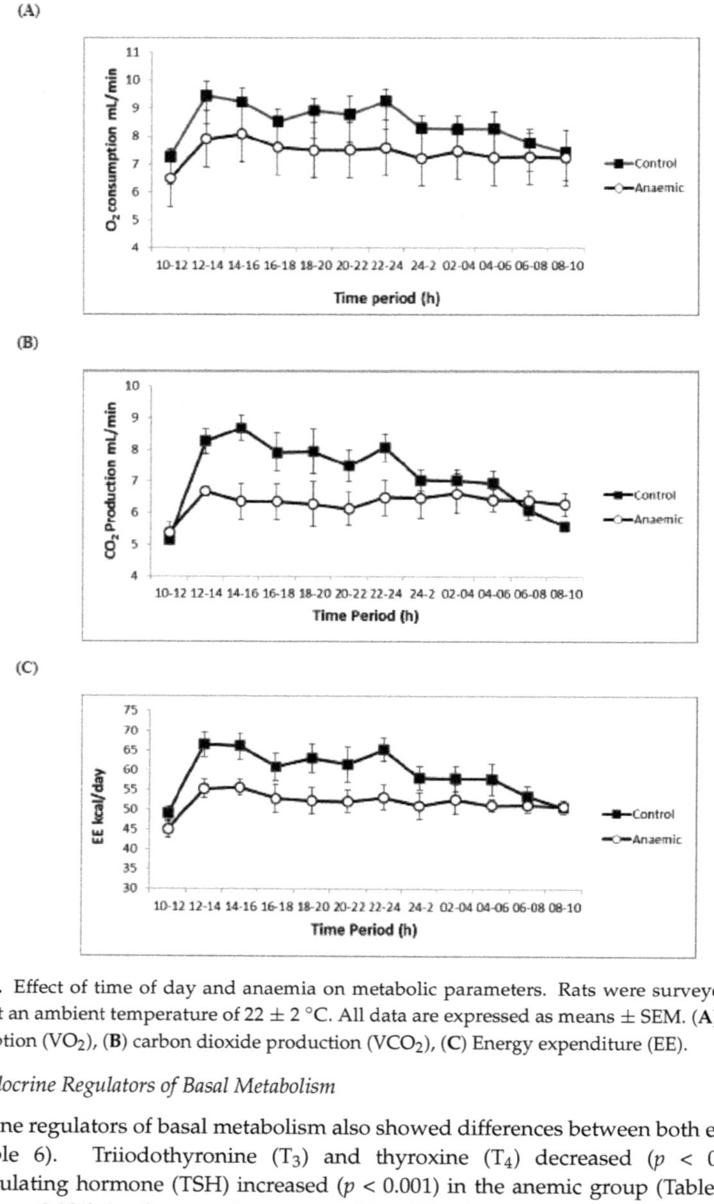

Figure 1. Effect of time of day and anaemia on metabolic parameters. Rats were surveyed every 10 min at an ambient temperature of 22 ± 2 °C. All data are expressed as means ± SEM. (**A**) oxygen consumption (VO_2), (**B**) carbon dioxide production (VCO_2), (**C**) Energy expenditure (EE).

3.5. Neuroendocrine Regulators of Basal Metabolism

Endocrine regulators of basal metabolism also showed differences between both experimental groups (Table 6). Triiodothyronine (T_3) and thyroxine (T_4) decreased ($p < 0.001$) while thyroid-stimulating hormone (TSH) increased ($p < 0.001$) in the anemic group (Table 6). Ghrelin decreased ($p < 0.001$) in the anemic group, while GIP, glucagon, insulin, corticosterone and adrenocorticotropic hormone (ACTH) increased in the Fe-deficient group ($p < 0.01$ for GIP, glucagon and insulin; $p < 0.001$ for corticosterone and ACTH.

Table 6. Plasma concentration of hormones affecting the basal metabolic rate.

	Normal-Fe Control Group (n = 20)	Low-Fe Anaemic Group (n = 20)
TSH (pg/mL)	29.43± 3.96	51.32 ± 5.21 **
T_3 (pg/mL)	15,269.21 ± 432.21	10,123.29 ± 348.17 **
T_4 (pg/mL)	1422.45 ± 98.77	1098.32 ± 95.37 **
Ghrelin (pg/mL)	36.42 ± 1.05	14.45 ± 0.81 **
GIP (pg/mL)	26.98 ± 1.24	33.28 ± 1.65 *
Glucagon (pg/mL)	24.23 ± 1.13	30.18 ± 1.24 *
Insulin (pg/mL)	726.43 ± 32.18	862.22 ± 29.98 *
Corticosterone (ng/mL)	177.65 ± 27.45	341.65 ± 41.24 **
ACTH (pg/mL)	1252.11± 109.23	861.32± 87.54 **

* Significantly different ($p < 0.01$) from the control group by Student's t-test. ** Significantly different ($p < 0.001$) from the control group by Student's t-test. TSH Thyroid-stimulant hormone, GIP Gastric inhibitory polypeptide, ACTH Adrenocorticotropic hormone.

4. Discussion

The study investigated changes in basal metabolism hormones, body composition, and energy expenditure during severe iron deficiency anaemia in young rats. Transferrin saturation, the most widely used screening measurement in Fe-deficiency anaemia, is low in this pathology [12,13], resulting in agreement with those obtained in the current study. MCV is a key red-cell marker for detecting Fe-deficiency anaemia in erythrocytes [14]. Serum ferritin has been a routine laboratory measure of Fe status because it is a well-standardized measurement for identifying Fe-deficiency because its concentration is directly proportional to body Fe stores [15]. On the other hand, RDW increased significantly in IDA because circulating Fe did not reach the bone marrow and red cells were deprived from Fe until this size was elevated. In IDA, Fe is not present in the enterocytes and macrophages because it is pumped from enterocytes across ferroportin 1 (FPN1) channel because hepcidin is low and therefore unable to prevent Fe releasing into the plasma [16]. Therefore, in light of all that is mentioned above, we can report that the body stores were significantly depleted after dietary Fe restriction during 40 days.

Bilirubin is the terminal product of heme metabolism, and total serum bilirubin levels are increased in hepatocellular injury, intra and extra hepatic biliary duct obstructions, intravascular and extravascular hemolysis. In the current study, bilirubin was increased and AST and ALT activities were also found elevated in IDA, which suggest necrosis liver function stress and impairment [17]. In our study, Fe-deficiency increased cortisol secretion, a fact that would increase energy expenditure and suppress body fat accumulation [12]. This increase in cortisol secretion is also associated with enhanced lipolysis, plasma triglyceride and cholesterol [18], supporting the changes in body composition of the anaemic group. With regard to glucose, increase in the Fe-deficient group, this can be attributed to the hyperglycemic effect of cortisol.

The differences in the on anaemic group reveals that hepatic Fe content is depleted and a severe degree of Fe-deficiency has been reached, induced by consuming a low-Fe diet for 40 days. The body weight is lower in anaemic animals, and hepatic weight also descended slightly. As a consequence, the hepatosomatic index (ratio liver weight/ body weight) is higher in the Fe-deficient animals. The lower Fe content recorded in liver in situation of IDA can be due to the decrease of the hepcidin recorded in the current study, a fact that is in agreement with the results reported previously by [19]. Hepcidin is a hepatic peptide, which has been identified as the master regulator of Fe metabolism [20]. Studies of Fe deficiency in hepcidin KO mice suggest that this peptide intervenes in regulating the storage of the mineral (attenuating both the intestinal Fe absorption and the liberation of the Fe of the macrophages) [21].

On the other hand, a low expression of hepcidin during Fe overload suggests that this protein is also a key factor in the erythroid regulation [22]. The diminished expression of hepcidin in response to the hypoxia induced by the Fe-deficiency, and its increased expression in mice and humans suffering with inflammation processes suggests that this hormone also takes part in the hypoxia homeostatic response and in the pathways of inflammatory regulators [23]. Hepcidin also controls the Fe levels directly acting reciprocally with the FPN1, driving to the internalization and degradation of FPN1 when levels of Fe are high, consequently blocking the liberation of the Fe in the sites of storage: hepatocytes, enterocytes and macrophages [24].

During Fe-deficiency, several regulatory factors of the hepcidin are impaired (the erythropoietic demand increases due to the decrease of the hematologic parameters, there is a minor supply of oxygen to the tissues and the body stores are depleted, being the Fe metabolism altered), facts that lead to a decrease of the above mentioned hormone, which will be translated in a minor interaction with the FPN1 [25], avoiding its internalization and degradation, therefore increasing the outflow of Fe^{2+} from the hepatocytes and its storage diminishes consistently in the above mentioned organ.

On the other hand, an Fe-restrictive diet causes a severe iron deficient anaemia enough to impair the body weight, which results in agreement with those reported by other authors [26], and this can be attributed to the lower levels of thyroid hormones found in this pathology in the current study and in previous reports [27]. Fe deficiency–induced alterations in the central nervous system impair thermoregulatory responses that induce the lower thyroid hormone response and the overall failure to thermoregulate that characterize Fe-deficient rats, diminishing basal metabolic rate and weight gain. TSH increased as a compensatory mechanism to enhance T_3 and T_4 production, diminished by the Fe-deficiency. In addition, our group reported that severe IDA has a significant impact upon bone turnover and mass, decreasing bone formation and increasing its resorption [28], thus implying delayed skeletal development. This can also explain reduced weight. Additionally, impairment of metabolic energy output, and cellular respiration of rats caused by Fe deficiency could be also responsible for the reduced body weight.

After Fe deprivation in the anemic group, cortisol levels increased significantly (Table 2). This can possibly be explained by the lower activity of monoaminooxidase in IDA, and the low level of aldehyde oxidase activity. These enzymatic changes induce an increase in the endogenous levels of circulating catecholamines, which enhance the release rate of adrenocorticotrophic hormone, producing in turn an increase in glucocorticoids output [12]. In addition, anemia can be considered a pro-inflammatory state, arising in part from a defect in the normal compensatory production of erythropoietin in response to a declining hemoglobin concentration [29]. In addition, inflammatory cytokines may negatively influence Fe absorption and recycling, thereby interfering with hemoglobin synthesis [30]. Therefore, the increase in corticosterone could be a compensatory mechanism to alleviate, at least partly, the anemia-induced inflammatory state because Glucocorticoids protect endothelial and epithelial cells from stress-induced apoptosis, exhibiting potent anti-inflammatory effects [31].

Ghrelin levels were decreased in anemia situation. In the presence of catabolism, plasma total ghrelin levels are increased, suggesting that, in that case, ghrelin does not increase food intake and/or anabolism under these circumstances. In addition, it is currently unknown whether administration of additional ghrelin in these conditions may reduce the development of cachexia [32]. Ghrelin levels detected in the current study decreased in parallel to Fe stores. Therefore, there is a decrease in appetite during IDA, leading to a reduction in body weight of the rats. The reduced ghrelin secretion could be a compensatory mechanism in reaction to the metabolic consequences of increased corticosterone secretion (Table 2) and induced catabolism [33].

Insulin was higher in the IDA group, together with an increase in glucagon, which could be due to the increased corticosterone secretion. The increase in insulin production has been previously reported in IDA [34]. These authors reported that, although basal glucose was significantly elevated in rats with IDA, there was also an increase in peripheral insulin responsiveness, a fact that could be beneficial in keeping glucose levels within a physiological range, in spite of the high corticosterone secretion. However, in spite the increased insulin responsiveness which could be beneficial to the higher insulin and high glucagon output, an increase in basal glucose was observed (Table 2).

Upon dietary ingestion, glucose-dependent insulinotropic polypeptide (GIP) hormones are secreted by K cells from the upper small intestine and they induce pancreatic cells to release insulin [35]. Consequently, the higher insulin rate secretion found in the current study can be explained by an increase in GIP secretion. In addition, studies in rodent models indicate that regulation of fat metabolism is an important physiological function of GIP that induces glucose uptake, activity of lipoprotein lipase, and accumulation of triglycerides by adipocytes [36], explaining the increase in total cholesterol and triglycerides found in the anemic rats (Table 2).

There are controversies in several studies, which reported associations between fat mass and various markers of Fe status [2,6–8,37,38]. In these studies, the direction of the association depends on the Fe biomarker (inverse for serum Fe, positive for serum ferritin) [2,37,38]. The report of Yanoff [38], indicating an association between obesity and serum ferritin concentration, suggests that the degree of fat mass might play a major role in predicting body Fe status. In other words, an association might be observed in subjects with a higher percent body fat but not in those in the normal range of adiposity, as occurs in the current study in which the animals are non-obese. Based on the study by Yanoff et al. [38], the hypoferremia observed during obesity appears to be explained both by true Fe deficiency and by inflammatory-mediated functional Fe deficiency.

In the present study, decreases in the body weight and adipose tissue mass were recorded in Fe-deficient rats, without changing the energy intake. In terms of energy intake, cortisol secreted during sympathetic nervous system action plays a key role. It is well known that normal digestion and absorption of dietary fat requires the action of pancreatic lipase. Because the sympathetic nervous system regulates, to a large extent, thermogenesis and fat oxidation, and cortisol release provides an explanation to a fat mass management in the situation of anemia [39,40].

Another important finding in this study was that body composition analysis showed increased free water in Fe-deficient animals, indicating edema and meaning a severe hemodynamic alteration together with overload of rennin–angiotensin–aldosterone system, which is unable to maintain the distribution of water between the intra and extracellular spaces [41].

With regard to the low O_2 consumption, CO_2 production and EE in the anaemic group, since in an Fe-deficiency situation, the Hb and RBC count drastically diminishes the supply of O_2 to the cells, which limits itself considerably. The lower O_2 consumption recorded in the current study affects in a negative way to the ATP synthesis, limiting and reducing the EE and impairing weight gain. These results agree with the findings of Schneider et al. [42], who reported that low levels of Hb significantly harm the weight gain, explaining therefore the association found in the present study between the minor weight and the Fe-deficiency. A limitation of the study is that statistical power could be impaired by fewer numbers (20 for each group), although, in many similar studies, the numbers were even lower (10 for each group), and numbers could be increased in future studies. Another limitation of the study is that, although animals are supplied healthy and pathongens-free, markers for infection and inflammation were not determined in serum and tissue samples of the mice used in the study.

5. Conclusions

In conclusion, the severe induced Fe-deficiency provokes an impairment of haematological and biochemical parameters and a depletion of the hepatic Fe levels together with a slight decrease of the liver size that negatively impacts the weight gain of animals in growth, since the hypoxia induced by the lack of Fe limits ATP production. Fe-deficiency impairs body weight and it is possibly due to the lower levels of thyroid hormones also observed in the current study. The increase in TSH recorded would be a compensatory mechanism to induce T_3 and T_4 production, diminished by the Fe-deficiency. After Fe deprivation in the anemic group, an increase in the cortisol levels was also recorded. Ghrelin levels decreased in an anemic situation, explaining the cachexia induced by Fe-deficiency and leading to reductions of lean mass. GIP increased during Fe deficiency leading to a higher insulin rate secretion. In addition, IDA showed marked reductions lean mass and reduced body fat, indicating reduced energy stores. Body weight is a composite index of both the lean and fat mass, and it cannot totally account for the growth deficit in rats suffering from Fe-deficiency because of the difference in relative proportions of lean and fat mass in the rat.

Author Contributions: I.L.-A. designed the study. J.M.-F. and J.D.-C. performed the experiments, analyzed the data and wrote the manuscript. M.J.M.A. contributed to data analysis. All the authors approved the final version of the manuscript and agree to be accountable for all aspects of the work in ensuring that questions related to the accuracy or integrity of any part of the work are appropriately investigated and resolved. All persons designated as authors qualify for authorship, and all those who qualify for authorship are listed.

Funding: This study was supported by the Excellence Project (P11-AGR-7648) from the Regional Government of Andalusia.

Acknowledgments: J.M.-F. was supported by fellowship from the Ministry of Education, Culture and Sport (Spain) and is grateful to the Excellence Ph.D. Program "Nutrición y Ciencias de los Alimentos" from the University of Granada. The authors are grateful to Susan Stevenson for her efficient support in the revision with the English language.

Conflicts of Interest: The authors declare no conflict of interest.

References

1. DeLoughery, T.G. Microcytic anemia. *N. Engl. J. Med.* **2014**, *371*, 1324–1331. [CrossRef] [PubMed]
2. Chambers, E.C.; Heshka, S.; Gallagher, D.; Wang, J.; Pi-Sunyer, F.X.; Pierson, R.N., Jr. Serum iron and body fat distribution in a multiethnic cohort of adults living in New York City. *J. Am. Diet. Assoc.* **2006**, *106*, 680–684. [CrossRef] [PubMed]
3. Zafon, C.; Lecube, A.; Simo, R. Iron in obesity. An ancient micronutrient for a modern disease. *Obes. Rev.* **2010**, *11*, 322–328. [CrossRef] [PubMed]
4. Hurrell, R.F. Bioavailability of iron. *Eur. J. Clin. Nutr.* **1997**, *51*, S4–S8. [PubMed]
5. Popovic, V.; Duntas, L.H. Brain somatic cross-talk: Ghrelin, leptin and ultimate challengers of obesity. *Nutr. Neurosci.* **2005**, *8*, 1–5. [CrossRef] [PubMed]
6. Aderibigbe, O.R.; Pisa, P.T.; Vorster, H.H.; Kruger, S.H. The relationship between iron status and adiposity in women from developing countries: A review. *Crit. Rev. Food. Sci. Nutr.* **2014**, *54*, 553–560. [CrossRef]
7. Park, C.Y.; Chung, J.; Koo, K.O.; Kim, M.S.; Han, S.N. Hepatic iron storage is related to body adiposity and hepatic inflammation. *Nutr. Metab. (Lond.)* **2017**, 14:14. [CrossRef] [PubMed]
8. Stankowiak-Kulpa, H.; Kargulewicz, A.; Styszyński, A.; Swora-Cwynar, E.; Grzymisławski, M. Iron status in obese women. *Ann. Agric. Environ. Med.* **2017**, *24*, 587–591. [CrossRef] [PubMed]
9. Reeves, P.G.; Nielsen, F.H.; Fahey, G.C., Jr. AIN-93 purified diets for laboratory rodents: Final report of the American Institute of Nutrition ad hoc writing committee on the reformulation of the AIN-76A rodent diet. *J. Nutr.* **1993**, *123*, 1939–1951. [CrossRef]
10. Pallarés, I.; Lisbona, F.; López-Aliaga, I.; Barrionuevo, M.; Alférez, M.J.M. Effects of iron deficiency on the digestive utilization of iron, phosphorus, calcium and magnesium in rats. *Br. J. Nutr.* **1993**, *70*, 609–620. [CrossRef]
11. Tang, H.; Vasselli, J.; Wu, E.; Gallagher, D. In vivo determination of body composition of rats using magnetic resonance imaging. *Ann. N. Y. Acad. Sci.* **2000**, *904*, 32–41. [CrossRef] [PubMed]
12. Campos, M.S.; Barrionuevo, M.; Alferez, M.J.; Gomez-Ayala, A.E.; Rodriguez-Matas, M.C.; Lopez Aliaga, I.; Lisbona, F. Interactions among iron, calcium, phosphorus and magnesium in the nutritionally iron-deficient rat. *Exp. Physiol.* **1998**, *83*, 771–781. [CrossRef]
13. Thomas, C.; Thomas, L. Biochemical markers and hematologic indices in the diagnosis of functional iron deficiency. *Clin. Chem.* **2002**, *48*, 1066–1076. [PubMed]
14. Cook, J.D. Diagnosis and management of iron-deficiency anaemia. *Best Pract. Res. Clin. Haematol.* **2005**, *18*, 319–332. [CrossRef]
15. Finch, C.A.; Bellotti, V.; Stray, S.; Lipschitz, D.A.; Cook, J.D.; Pippard, M.J.; Huebers, H.A. Plasma ferritin determination as a diagnostic tool. *West. J. Med.* **1986**, *145*, 657–663. [PubMed]
16. Pasricha, S.R.; McQuilten, Z.; Westerman, M.; Keller, A.; Nemeth, E.; Ganz, T.; Wood, E. Serum hepcidin as a diagnostic test of iron deficiency in premenopausal female blood donors. *Haematologica* **2011**, *96*, 1099–1105. [CrossRef]

17. Tolman, K.G.; Re, J.R. Liver function. In *Tietz Textbook of Clinical Chemistry*; Burtis, C.A., Ashwood, E.R., Eds.; Saunders: Philadelpia, PA, USA, 1999; pp. 1125–1177.
18. Ikeuchi, M.; Koyama, T.; Takahashi, J.; Yazawa, K. Effects of astaxanthin in obese mice fed a high-fat diet. *Biosci. Biotechnol. Biochem.* **2007**, *71*, 893–899. [CrossRef]
19. Theurl, I.; Aigner, E.; Theurl, M.; Nairz, M.; Seifert, M.; Schroll, A.; Sonnweber, T.; Eberwein, L.; Witcher, D.R.; Murphy, A.T.; et al. Regulation of iron homeostasis in anemia of chronic disease and iron deficiency anemia: Diagnostic and therapeutic implications. *Blood* **2009**, *113*, 5277–5286. [CrossRef]
20. Viatte, L.; Vaulont, S. Hepcidin, the iron watcher. *Biochimie* **2009**, *91*, 1223–1228. [CrossRef]
21. Finberg, K.E. Striking the target in iron overload disorders. *J. Clin. Investig.* **2013**, *123*, 1424–1427. [CrossRef]
22. Weinstein, D.A.; Roy, C.N.; Fleming, M.D.; Loda, M.F.; Wolfsdorf, J.I.; Andrews, N.C. Inappropriate expression of hepcidin is associated with iron refractory anemia: Implications for the anemia of chronic disease. *Blood* **2002**, *100*, 3776–3781. [CrossRef] [PubMed]
23. Nicolas, G.; Chauvet, C.; Viatte, L.; Danan, J.L.; Bigard, X.; Devaux, I.; Beaumont, C.; Kahn, A.; Vaulont, S. The gene encoding the iron regulatory peptide hepcidinis regulated by anemia, hypoxia, and inflammation. *J. Clin. Investig.* **2002**, *110*, 1037–1044. [CrossRef]
24. Nemeth, E.; Ganz, T. Regulation of iron metabolism by hepcidin. *Annu. Rev. Nutr.* **2006**, *26*, 323–342. [CrossRef]
25. Tselepis, C.; Ford, S.J.; McKie, A.T.; Vogel, W.; Zoller, H.; Simpson, R.J.; Diaz Castro, J.; Iqbal, T.H.; Ward, D.G. Characterization of the transition-metal-binding properties of hepcidin. *Biochem. J.* **2010**, *427*, 289–296. [CrossRef] [PubMed]
26. Strube, Y.N.; Beard, J.L.; Ross, A.C. Iron deficiency and marginal vitamin A deficiency affect growth, hematological indices and the regulation of iron metabolism genes in rats. *J. Nutr.* **2002**, *132*, 3607–3615. [CrossRef] [PubMed]
27. Beard, J.L.; Brigham, D.E.; Kelley, S.K.; Green, M.H. Plasma thyroid hormone kinetics are altered in iron-deficient rats. *J. Nutr.* **1998**, *128*, 1401–1408. [CrossRef] [PubMed]
28. Diaz-Castro, J.; Lopez-Frias, M.R.; Campos, M.S.; Lopez-Frias, M.; Alferez, M.J.; Nestares, T.; Ojeda, M.L.; Lopez-Aliaga, I. Severe nutritional iron-deficiency anaemia has a negative effect on some bone turnover biomarkers in rats. *Eur. J. Nutr.* **2012**, *51*, 241–247. [CrossRef]
29. Wilson, A.; Yu, H.T.; Goodnough, L.T.; Nissenson, A.R. Prevalence and outcomes of anemia in rheumatoid arthritis: A systematic review of the literature. *Am. J. Med.* **2004**, *116*, 50S–57S. [CrossRef] [PubMed]
30. Nemeth, E.; Rivera, S.; Gabayan, V.; Keller, C.; Taudorf, S.; Pedersen, B.K.; Ganz, T. IL-6 mediates hypoferremia of inflammation by inducing the synthesis of the iron regulatory hormone hepcidin. *J. Clin. Investig.* **2004**, *113*, 1271–1276. [CrossRef]
31. Newton, C.J.; Ran, G.; Xie, Y.X.; Bilko, D.; Burgoyne, C.H.; Adams, I.; Abidia, A.; McCollum, P.T.; Atkin, S.L. Statin-induced apoptosis of vascular endothelial cells is blocked by dexamethasone. *J. Endocrinol.* **2002**, *174*, 7–16. [CrossRef]
32. Akarsu, S.; Ustundag, B.; Gurgoze, M.K.; Sen, Y.; Aygun, A.D. Plasma ghrelin levels in various stages of development of iron deficiency anemia. *J. Pediatr. Hematol. Oncol.* **2007**, *29*, 384–387. [CrossRef]
33. Otto, B.; Tschop, M.; Heldwein, W.; Pfeiffer, A.F.; Diederich, S. Endogenous and exogenous glucocorticoids decrease plasma ghrelin in humans. *Eur. J. Endocrinol.* **2004**, *151*, 113–117. [CrossRef] [PubMed]
34. Borel, M.J.; Beard, J.L.; Farrell, P.A. Hepatic glucose production and insulin sensitivity and responsiveness in iron-deficient anemic rats. *Am. J. Physiol.* **1993**, *264*, 380–390. [CrossRef] [PubMed]
35. Miyawaki, K.; Yamada, Y.; Yano, H.; Niwa, H.; Ban, N.; Ihara, Y.; Kubota, A.; Fujimoto, S.; Kajikawa, M.; Kuroe, A.; et al. Glucose intolerance caused by a defect in the entero-insular axis: A study in gastric inhibitory polypeptide receptor knockout mice. *Proc. Natl. Acad. Sci. USA* **1999**, *96*, 14843–14847. [CrossRef] [PubMed]
36. Gault, V.A.; McClean, P.L.; Cassidy, R.S.; Irwin, N.; Flatt, P.R. Chemical gastric inhibitory polypeptide receptor antagonism protects against obesity, insulin resistance, glucose intolerance and associated disturbances in mice fed high-fat and cafeteria diets. *Diabetologia* **2007**, *50*, 1752–1762. [CrossRef] [PubMed]
37. Iwasaki, T.; Nakajima, A.; Yoneda, M.; Yamada, Y.; Mukasa, K.; Fujita, K.; Fujisawa, N.; Wada, K.; Terauchi, Y. Serum ferritin is associated with visceral fat area and subcutaneous fat area. *Diabetes Care* **2005**, *28*, 2486–2491. [CrossRef] [PubMed]
38. Yanoff, L.B.; Menzie, C.M.; Denkinger, B.; Sebring, N.G.; McHugh, T.; Remaley, A.T.; Yanovski, J.A. Inflammation and iron deficiency in the hypoferremia of obesity. *Int. J. Obes.* **2007**, *31*, 1412–1419. [CrossRef]

39. Landsberg, L.; Young, J.B. Sympathoadrenal activity and obesity: Physiological rationale for the use of adrenergic thermogenic drugs. *Int. J. Obes. Relat. Metab. Disord.* **1993**, *17*, S29–S34.
40. Arch, J.R.; Wilson, S. Prospects for beta 3-adrenoceptor agonists in the treatment of obesity and diabetes. *Int. J. Obes. Relat. Metab. Disord.* **1996**, *20*, 191–199.
41. Colin-Ramirez, E.; Castillo-Martinez, L.; Orea-Tejeda, A.; AsensioLafuente, E.; Torres Villanueva, F.; Rebollar Gonzalez, V.; Narvaez David, R.; Dorantes Garcia, J. Body composition and echocardiographic abnormalities associated to anemia and volume overload in heart failure patients. *Clin. Nutr.* **2006**, *25*, 746–757. [CrossRef]
42. Schneider, J.M.; Fujii, M.L.; Lamp, C.L.; Lönnerdal, B.; Dewey, K.G. The use of multiple logistic regression to identify risk factors associated with anemia and iron deficiency in a convenience sample of 12–36-mo-old children from low-income families. *Am. J. Clin. Nutr.* **2008**, *87*, 614–620. [CrossRef] [PubMed]

nutrients MDPI

Review

Iron Deficiency and Iron Homeostasis in Low Birth Weight Preterm Infants: A Systematic Review

Jorge Moreno-Fernandez [1,2], Julio J. Ochoa [1,2,*], Gladys O. Latunde-Dada [3] and Javier Diaz-Castro [1,2]

[1] Department of Physiology, University of Granada, 18071 Granada, Spain; jorgemf@ugr.es (J.M.-F.); javierdc@ugr.es (J.D.-C.)
[2] Institute of Nutrition and Food Technology "José Mataix Verdú", University of Granada, 18071 Granada, Spain
[3] Department of Nutritional Sciences, School of Life Course Sciences, Faculty of Life Sciences & Medicine, King's College London, London SE1 9NH, UK; yemisi.latunde-dada@kcl.ac.uk
* Correspondence: jjoh@ugr.es; Tel.: +34-958241000 (ext. 20317)

Received: 11 February 2019; Accepted: 15 May 2019; Published: 16 May 2019

Abstract: Iron is an essential micronutrient that is involved in many functions in humans, as it plays a critical role in the growth and development of the central nervous system, among others. Premature and low birth weight infants have higher iron requirements due to increased postnatal growth compared to that of term infants and are, therefore, susceptible to a higher risk of developing iron deficiency or iron deficiency anemia. Notwithstanding, excess iron could affect organ development during the postnatal period, particularly in premature infants that have an immature and undeveloped antioxidant system. It is important, therefore, to perform a review and analyze the effects of iron status on the growth of premature infants. This is a transversal descriptive study of retrieved reports in the scientific literature by a systematic technique. PRISMA (Preferred Reporting Items for Systematic Reviews and Meta-Analyses) guidelines were adapted for the review strategy. The inclusion criteria for the studies were made using the PICO (population, intervention, comparison, outcome) model. Consequently, the systematic reviews that included studies published between 2008–2018 were evaluated based on the impact of iron status on parameters of growth and development in preterm infants.

Keywords: iron; growth; development; infant; premature

---

1. Introduction

Iron is involved in many key cellular functions and processes in humans due to its role as an essential micronutrient. This ubiquitous mineral plays a critical role in growth and the development of the central nervous system. It is essential for energy metabolism, cell differentiation, and a host range of physiological processes for the normal functioning of the brain [1]. In spite of this, iron deficiency (ID) is the most prevalent micronutrient deficiency in the world, and it affects all age groups, children between 0 and 5 years of age being the most affected [2]. Therefore, iron requirements are particularly high during the periods of rapid growth and differentiation and imbalance in iron homeostasis could result in alterations of cognitive functions and neurodevelopment [3].

Preterm infants are prone to develop iron deficiency anemia (IDA) in the first 4 months of life due to lower iron stores at birth compared with term infants, rapid growth and iron losses. Most fetal iron is transferred from the mother during the third trimester of gestation. This transfer is interrupted by preterm birth, resulting in iron stores at birth being proportional to birth weight. Despite low iron stores at birth, growth velocity of premature infants is maximal at a postmenstrual age of 28–38 weeks, reflecting a particularly high iron need. The risk of IDA in preterm infants is further increased by frequent uncompensated iatrogenic phlebotomy losses [3–5].

Initial evaluation should include a clinical history, to assess prematurity, low birth weight, diet, chronic diseases, family history of anemia, and ethnic background. A complete blood count is the most common initial diagnostic test used to evaluate for IDA, and this count allows for differentiating microcytic, normocytic, and macrocytic anemia based on the mean corpuscular volume (MCV). Ferritin measurement is the most sensitive test for diagnosing IDA, being a good reflection of total iron storage and is also the first laboratory index to decline with ID [6].

Developmental outcomes are positively influenced by adequate nutrition early in life, revealing an improvement in neurodevelopment and cognitive abilities in preterm infants, for instance. During development, ID negatively affects the growth and functioning of multiple organ systems particularly the brain, skeletal muscle, heart and the gastrointestinal tract [7–11]. IDA in preterm infants could induce impaired cell differentiation [12,13] and alter normal neurodevelopment processes [14,15]. Due to increased risk of developing IDA in premature infants, iron supplementation is recommended in some situations. Although iron supplementation improves iron status of preterm infants [16], the nature and the means by which this is achieved are not clearly defined [17,18]. Preterm infants do not have a fully developed antioxidant capacity and free iron acts as potential oxidative stressor, iron supplementation, therefore, has to be considered cautiously in this population [4].

An excess of iron can affect organ development during the postnatal period. There is an association between iron overload, retinopathy of prematurity, and bronchopulmonary dysplasia. Non-protein bound iron mediates oxidative stress in the presence of poor antioxidant capabilities, and this condition initiates or potentiates the progression of these pathologies through generation of free radicals [4]. ID and toxicity could be avoided by an adequate iron homeostasis, which is essential for optimal development and function.

There is a high prevalence of preterm babies who have an increased risk of IDA, and there are no clinical consensuses or guidelines about the clinical benefits of iron supplementation and/or fortification that the hospitals should follow with this extremely vulnerable population. Therefore, it is relevant to evaluate the effects of ID or iron toxicity on the optimal development and function of the organs such as the brain in premature infants, ID and iron overload disorders during this period [4].

Consequently, this systematic review performed an analysis of the influence and effects of iron on the hematological parameters, growth, and development of premature infants. Thus, the review also revealed a systematic literature evidence on the beneficial effects of iron supplementation in premature infants.

2. Materials and Methods

2.1. Design

Descriptive study of the articles retrieved using a systematic technique.

2.2. Sources of Data Extraction

The data used in this study were obtained from the consultation and direct access to the scientific literature collection in the following databases: Medlars International Literature Online (MEDLINE) via PubMed (USA National Library of Medicine National Institutes of Health Search database), The Cochrane Library Plus, SCOPUS (World's largest abstract and citation database of peer-reviewed literature), Web of Science and Literatura Latinoamericana y del Caribe en Ciencias de la Salud (LILACS).

2.3. Search Strategy

A search strategy for electronic databases was developed with regard to information processing, from the Thesaurus study, Medical Subject Headings (MeSH) developed by the U.S. National Library of Medicine.

The use of the following terms was considered appropriate: "Iron", "Growth and Development", and "Infant, Premature", as well as descriptors as text in the title and summary record fields. A search equation was developed for use in the MEDLINE database, via PubMed, using boolean connectors, and then adapted to the rest of aforementioned data bases: ("Iron"(Mesh) AND "Growth and Development"(Mesh) AND "Infant, Premature"(Mesh)). The search was expanded to include "Anemia, Iron-Deficiency" (Mesh).

2.4. Eligibility Criteria

The proposed criteria for the studies included in the review, considering the PICO (population, intervention, comparison, outcome) model [19] were: (a) studies published in peer-reviewed journals; (b) studies involving infants who were premature (gestational age <37 weeks) or with a low birth weight (<2500 g); they are especially susceptible to developing IDA because these infants have smaller iron stores at birth and a greater iron requirement concurrent with a rapid increase in the red cell mass; (c) studies had to inform quantitatively about the iron status and iron supplementation.

Studies were excluded if the full text was not found, if they were not carried out in humans, if the premature infants presented diseases different to anemia, if they were identified as redundant publications, if they were not written in English or Spanish, or if they did not include an empirical result related to the iron status in the organism of premature infants and their influence on growth and development. We also limited references to the last 10 years (2008–2018), as the calculated period of obsolescence (half period of Burton Kebler [20]).

A multidisciplinary research team developed the search, inclusion and exclusion criteria. Pediatrics and nutritionist research experts also took part in this process. Furthermore, a biostatistician was involved in this phase.

2.5. Study Selection

The initial phase of criteria selection was the exclusion of duplicates. Articles whose titles and/or abstracts were not clearly related to the subject of study were also excluded. Once all potentially eligible articles had been located, each of them was carefully reviewed to establish further eligibility criteria and all relevant information and data were extracted.

The selection of the articles was performed independently by two authors: J.M-F and J.D-C, and it was established that the valuation of the concordance between the authors (Kappa index) should be greater than 80% [21]. In the case of a discrepancy, the authors reviewing the literature made a consensus decision. Additionally, the reference list of every selected article was carefully checked to identify other potentially eligible articles, which were processed in the same way as those retrieved from the electronic search. Independently, both reviewers read the full text of the remaining articles and screened for inclusion. Afterwards, they compared the results of the full text screening. A predefined template was used to perform data extraction from eligible articles.

2.6. Data Extraction

Two independent reviewers extracted the data. To increase uniformity and reduce bias, a standardized data collection form was used by each reviewer. The reviewers made a consensus decision when there was a discrepancy. From each eligible study, the following information/data were extracted: first author, year of publication, journal, eligibility criteria, definition of premature or low-birth weight, number of participants, hematological data (hemoglobin (Hb), hematocrit (Ht), serum iron) before iron supplementation, iron supplement dose, duration of iron supplementation, hematological data after iron supplementation, growth status after iron supplementation, adverse effects of iron supplementation, and effects of iron on cognitive status.

2.7. Risk of Bias in Individual Studies

The final sample of studies for review was subsequently assessed by the authors (Table 1) separately to convey the methodological and critical appraisal using the Delphi List appraisal tool [22] for randomized studies. This tool is composed of an eight item scale, including questions that are rated according to yes/no/don't know. A response of "yes" is indicated by the numeric score of 1 and a response of "no" or "don't know" is indicated by a 0 (with a maximum score of 8). The Delphi List achieved consensus on a generic criteria list for quality assessment in randomized clinical trials. The adoption of this core set is the minimum reference standard of quality measures for all randomized clinical trials, where a score of 1 means high risk of bias and 8 means a low risk.

Non-randomized studies were assessed using the Transparent Reporting of Evaluations with Non-randomized Designs (TREND) statement [23], which contains a list of 22 essential aspects (available at http://www.TREND-statement.org) that must be described in the publication. The TREND statement is specifically developed to guide standardized reporting of nonrandomized controlled trials. The TREND statement complements the widely adopted CONsolidated Standards Of Reporting Trials (CONSORT) statement developed for randomized controlled trials. This is a collective effort in promoting the idea that transparent reporting is valuable to improve research synthesis and advance evidence-based recommendations for best practices and policies. A point was assigned for each item featured. Meanwhile, if an item was not applicable, it did not score (a maximum score of 22). When an item was composed of several points, these points were evaluated independently, giving the same value to each of them, and subsequently an average was made (being the final result of each item), in such a way that in no case the score of one point per item could be exceeded. For each included study, a consensus was achieved after discussion of discrepancies. Tables 1 and 2 summarize the score for each study and criterion included in the Delphi List and the TREND statement. Based on the quality appraisal, no studies were discarded.

Upon thoroughly reviewing the data collection and statistical analysis of each included article, it became apparent that there was too much variation between these studies, and, therefore, it was not considered appropriate to conduct additional analysis such as a meta-analysis using the reported findings.

J.M-F and J.D-C reviewed the articles independently, extracted relevant data from each included article, and assessed risk of bias. The Delphi List was carried out in the 5 randomized studies selected, while the TREND Checklist was used in the 9 non-randomized selected manuscripts. Any discrepancies in scoring were resolved by mutual discussion or through discussion with J.J.O. An overall risk of bias rating was assigned to each study based on the quality score.

The Delphi List carried out in the selected manuscripts is provided in Table 1. The total score for each report is shown in the last file constituting the sum of scores for each criterion. The maximum possible score was 8, but this was not assigned to any manuscript. Reports ranged between 5 [24,25] and 7 [26–29], with an average score of 6.2. The TREND Statement Checklist is provided in Table 2. The total score for each report is shown in the last file constituting the sum of scores for each criterion. The maximum possible score was 22 but was not assigned to any manuscript. Manuscripts ranged between 10.92 and 13.85, with an average score of 12.19.

Table 1. Score breakdown on 8 items from the Delphi list for each randomized study.

Study	Items from the Delphi List									
	1	2	3	4	5	6	7	8	Total Score	
Gupta et al., 2017 [25]	1	1	1	NA	NA	NA	1	1	5	
Berglund et al., 2018 [29]										
Berglund et al., 2013 [28]	1	1	1	1	NA	1	1	1	7	These 4 references are all results from the same original cohort
Berglund et al., 2011 [27]										
Berglund et al., 2010 [26]										
Sankar et al., 2009 [24]	1	1	1	0	NA	NA	1	1	5	

Items of Delphi List: 1. Treatment allocation; 2. Were the groups similar at baseline regarding the most important prognostic indicators?; 3. Were the eligibility criteria specified?; 4. Was the outcome assessor blinded?; 5. Was the care provider blinded?; 6. Was the patient blinded?; 7. Were point estimates and measures of variability presented for the primary outcome measures?; 8. Did the analysis include an intention-to-treat analysis?; "yes" is indicated by a numeric score of 1 and "no" or "don't know" is indicated by a 0; N/A: not applicable.

Table 2. Score breakdown on 22 items from the Transparent Reporting of Evaluations with Non-randomized Designs (TREND) Statement Checklist for each non-randomized study.

Study	Items from the TREND Statement Checklist																						
	1	2	3	4	5	6	7	8	9	10	11	12	13	14	15	16	17	18	19	20	21	22	Total Score
de Waal et al., 2017 [30]	1	1	0.75	N/A	1	0.66	0	0.33	1	0.5	0.25	0.85	0	0.25	0	0.5	0.5	0	0	0.33	1	1	10.92
Saha et al., 2016 [31]	1	1	0.5	N/A	1	0.66	1	0.66	0	0.5	0.75	0.85	1	0.5	0	1	0	1	0	0.33	1	1	13.75
Akkermans et al., 2016 [32]	1	1	1	N/A	1	0.66	0	0.33	0	0.5	0.75	0.71	0	0.5	0	0.5	0.5	1	0	0.33	1	1	11.78
Uijterschout et al., 2015 [33]	0.66	1	0.5	N/A	1	0.66	1	0.33	0	0.5	0.75	0.85	0	0.25	0	1	1	1	0	0.33	1	1	12.83
Yamada and Leone 2014 [34]	1	1	1	N/A	1	0.33	1	0.66	0	0.5	0.5	0.85	1	0.25	0	0.5	0.5	0	0	0.66	1	1	12.75
Mukhopadhyay et al., 2012 [35]	1	1	0.75	N/A	1	0.66	1	0	1	0.5	0.75	0.28	0	0.25	0	1	1	1	0	0.66	1	1	13.85
Amin et al., 2012 [36]	1	1	0.75	N/A	1	0.66	0	0.66	0	0.5	0.5	0.71	1	0.5	0	0.5	0.5	0	0	0.66	1	1	11.94
Kitajima et al., 2011 [37]	0.66	0.5	0.75	N/A	1	0.33	0	0	0	0	0.66	1	1	0.5	1	1	0.33	0	0	0.66	1	1	11.39
Molloy et al., 2009 [38]	0.33	1	0.75	N/A	1	0.66	1	0.66	0	0.5	0.5	0.85	1	0.5	0	1	0.33	0	0	0.33	1	1	12.41
Amin et al., 2010 [39]	1	1	0.75	N/A	1	0.33	0	0.33	0	0.5	0.5	0.71	0	0.5	0	1	0.33	0	0	0.33	1	1	10.28

N/A: not applicable.

3. Results

3.1. Study Selection

The above methodology resulted in a total description of search criteria and provided a total of 108 citations. The title and abstract of each report were reviewed by J.M.-F. and J.D.-C. to assess eligibility, and if the abstract or title was not described in enough detail, full reports were assessed to consider their inclusion. In this sense, all full-text reports for inclusion eligibility were independently assessed by J.M.-F. and J.D.-C. The level of agreement between the two researchers was high enough. The researchers resolved minor differences related to the inclusion of the reports by discussion. The potential lists of each relevant recovered article were scanned and four more articles were identified as possibly relevant.

Duplicate manuscripts and those based on exclusion criteria and limits were removed. Thirty studies remained potentially eligible (Figure 1). Sixteen studies were excluded because they did not meet the inclusion criteria. Finally, 16 studies met all inclusion criteria and were incorporated in the current review.

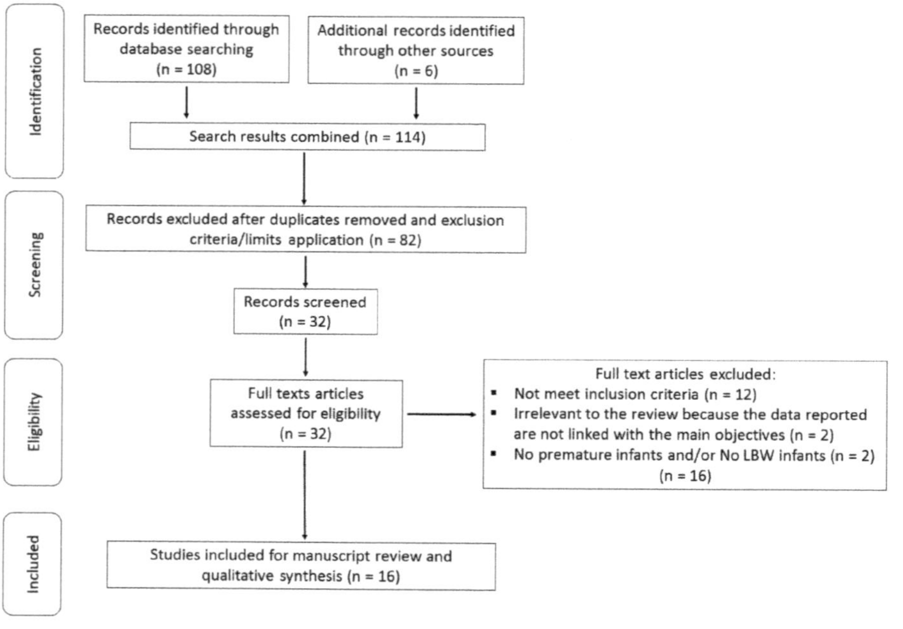

Figure 1. Flowchart showing study selection procedure and results.

3.2. Study Characteristics

The summary of the most relevant characteristics of the randomized and non-randomized studies included are shown in Tables 3 and 4, respectively. Out of the 16 studies included, four were conducted in India [24,25,31,35], three were carried out in the Netherlands [30,32,33], four were performed in Sweden [26–29], three in the USA [36,38,39], one in Japan [37], and one in Brazil [34]; these studies were published between 2009 and 2017.

Regarding the randomization, this review includes six randomized studies [24–29] (being [26–29] results from the same original cohort) and 10 non-randomized studies [30–39]. Four of these studies were double-blind [26–29].

Sample sizes ranged from 46 to 401 subjects. Among the total of 1743 participants included in the trials, 50.95% were men and 49.05% were women. One study did not specify the number of male and female participants [27]. The age preterm participants in the study ranged from 24 to 36 weeks. Yamada and Leone [34] included term newborns at 37 to 41 weeks of age. None of the included participants featured other pathologies apart from anemia. Most of the studies used iron supplements (specifically nine reports [24–29,34,36,38]) and the remaining studies did not use iron supplements [31–33,35,37,39].

Among all included studies, a variety of different iron status parameters were investigated. Most of the studies analyzed Hb, serum ferritin (SF), serum iron, and total iron-binding capacity (TIBC). Neurological development was also studied to measure the developmental assessment scale and child behavior checklist questionnaire. Berglund et al. [28] followed the participants from birth to 3.5 years of age and performed the Wechsler Preschool and Primary Scale of Intelligence (WPPSI-III) to assess cognitive function. The auditory brainstem response (ABR) was assessed as an indicator of the development of the central nervous system (CNS) [27,39], and a Developmental Assessment Scale for Indian Infants (DASII) was also used to evaluate neurodevelopment [25]. Moreover, Berglund 2018 tested children at 7 years of age for psychometric intelligence quotient (IQ) using the validated Wechsler Intelligence Scale for Children (fourth edition, WISC-IV), Child Behavior Checklist (CBCL) questionnaire, and Five to Fifteen (FTF) test.

Table 3. Summary of the characteristics of the randomized studies. LBW, low birth weight; CBCL, Child Behavior Checklist; ABR, auditory brainstem response; ID, iron deficiency.

Authors, Years	Study Design	Randomized	Sample Size	Male-Female	Preterm Weeks Age	Iron Supplementation	Dosage mg/kg/day	Key Findings
Berglund et al., 2018 [29]	controlled, double-blind, interventional trial	Yes	285	179–106	<37	Yes	1 or 2	Early iron supplementation in LBW infants prevents behavioral problems at school age, recommending iron supplementation in this population.
Gupta et al., 2017 [25]	open-label, multicentre trial	Yes	401	213–188	<34	Yes	2–3	There were no significant differences in motor and mental development quotients in premature infants who received complementary feeding at 4 months, or continuation of milk feeding and initiation of complementary feeding at 6 months.
Berglund et al., 2013 [28]	controlled, double-blind, interventional trial	Yes	319	158–161	<37	Yes	1 or 2	Iron supplementation reduced the prevalence of behavioral problems, defined as abnormal CBCL scores. Marginally low birth weight infants should be included in iron supplementation programs during early infancy.
Berglund et al., 2011 [27]	controlled, double blinded intervention trial	Yes	223	109–114	<37	Yes	1 or 2	Iron supplementation did not improve ABR latencies, and iron-deficient marginally low birth weight infants did not have impaired ABR latencies at 6 months. ABR is not a sensitive measure of impaired neurological development or that mild/moderate ID causes no such impairment in these infants.
Berglund et al., 2010 [26]	controlled, double-blind, interventional trial	Yes	285	138–147	<37	Yes	1 or 2	Iron supplementation from 6 weeks to 6 months reduces this risk of ID and IDA effectively, with no short-term adverse effects on morbidity or growth.
Sankar et al., 2009 [24]	Observacional	Yes	46	24–22	33	Yes	3–4	Supplementing iron at 2 weeks of life did not improve either serum ferritin or haematological parameters at 2 months of age in preterm very low birth weight infants.

Table 4. Summary of characteristics of non- randomized studies. ZnPP/H, Zinc protoporphyrin/heme; SF, serum ferritin.

Authors, Years	Study Design	Randomized	Sample Size	Men-Women	Preterm Weeks Age	Iron Supplementation	Dosage mg/kg/day	Key Findings
de Waal et al., 2017 [30]	prospective cohort study	No	161	98–63	32–36	No	–	ZnPP/H can be of additional value to detect infants at risk for IDA
Saha et al., 2016 [31]	prospective observational study	No	67	32–35	36.8–37.3	No	–	Late preterm and term small gestational age infants have adequate iron stores at birth and at 2 months of age
Akkermans et al., 2016 [32]	prospective multi-centre study	No	68	43–25	32–35	No	–	Iron depletion is common in late preterm infants at the age of 6 weeks in a setting without standardized iron supplementation. Early iron supplementation should be considered in late preterm infants with a low birth weight or low SF in the first week of life.
Uijterschout et al., 2015 [33]	prospective cohort study	No	143	87–56	32–36	No	–	Preterm infants have an increased risk of ID compared with those born at term. Supporting the need of iron supplementation. Measurement of ferritin at the age of 1 week might be useful to identify those infants at particular risk.
Yamada and Leone 2014 [34]	Cohort study	No	46	27–19	34–36	Yes	2	Exclusively breastfed late-preterm newborns presented greater reductions in hemoglobin/hematocrit and lower iron stores than term newborns. Specific iron supplementation is suggested.
Mukhopadhyay et al., 2012 [35]	prospective cohort study	No	150	47–103	≤36	No	–	Preterm small gestational age infants have lesser total body iron stores as compared to preterm adequate gestational age infants at birth. Similarly preterm infants have less iron stores than term infants.
Amin et al., 2012 [36]	prospective observational study	No	131	67–64	24–32	Yes	2	Iron parameters at 35 weeks post-menstrual age is extremely variable and is predicted by erythrocyte transfusions. Due to the harmful effects of iron overload and latent iron deficiency status, iron homeostasis maintenance is crucial in the neonatal period.
Kitajima et al., 2011 [37]	prospective cohort study	No	71	29–32	<37	No	-	Preterm infants have lower prohepcidin production at birth according to the gestational age, and the levels might be susceptible to the in utero stress. The postnatal increase might reflect the maturation and/or adaptation of iron homeostasis.
Amin et al., 2010 [39]	prospective cohort study	No	80	41–39	27–33	No	-	Premature infants with iron deficiency have abnormal auditory neural maturation compared with infants with normal iron status.
Molloy et al., 2009 [38]	Observational	No	60	22–38	26.5	Yes	2–4	Careful evaluation of iron indices is essential to prevent potential organ injury and unnecessary iron supplementation which could induce iron overload.

3.3. Characteristics of Iron Status Measurements

All of the studies included in the review assessed the iron status in preterm neonates to evaluate iron metabolism and requirements, supplying a large number of different measures. SF levels were reported in all the studies except Amin et al. [39], who measured cord ferritin. The SF level reflects total body iron stores, providing a convenient method of assessing the status of iron storage. SF is measured as a routine biomarker, although it is now known that many additional factors, including inflammation, infection, and malignancy complicate the interpretation of this value. Nine studies reported Hb levels [24,26–34] (Hb is a heterotetrameric protein that reversibly binds oxygen. Hb transports heme-bound oxygen from the lungs to all of the tissues in the body via oxygenation-linked shifts in the conformational equilibrium between the tense state and the relaxed state). This structural transition is controlled allosterically, as the binding of oxygen to one subunit affects the oxygen-affinity of the other subunits in the same tetrameric assembly), four measured Ht levels [24,31,34,38], four measured serum iron levels [26,29,34,35,38] (Ht is expressed as red blood cells packed cell volume, which is a very useful clinical factor in hemodialysis, surgical procedures, and anemia, which can be used to estimate transfusions and determine the extent of anemia), five measured transferrin saturation levels [26–29,34,37] (the transferrin saturation is an index that takes into account both plasma iron and its main transport protein and is considered an important biochemical marker of body iron status), three assayed TIBC levels [34,35,38] (TIBC indicates the maximum amount of iron necessary to saturate all available transferrin iron-binding sites, therefore, it correlates well with transferrin concentration), and three studies analyzed transferrin receptors [26–28] (the transferrin receptor mediates cellular iron uptake through clathrin-dependent endocytosis of iron-loaded transferrin, playing a key role in iron homeostasis).

Two studies measured reticulocytes counts [30,34], one study reported transferrin saturation [26,29], one manuscript revealed percentage of total iron binding capacity [38], and one study assessed red cell counts [37]. Moreover, one study incorporated Zinc protoporphyrin/heme (ZnPP/H) levels as a possible iron status indicator [30]. Berglund et al. [28] also measured hepcidin concentration. In addition, another study reported a serum prohepcidin concentration [37].

3.4. Iron Status and Body Composition

In all studies that analyzed the weight, the length and head circumference of premature infants was lower than that of the term infants. As suspected, these measurements were lower in preterm infants when compared to term infants, although Yamada and Leone [34] showed an increase in weight, length, head circumference, and BMI ($p < 0.001$) at an expected rate for the age of the population during the postnatal development of all infants, although these measurements were lower in preterm infants. The infants with iron depletion showed a lower birth weight than those without iron depletion. Lower birth weight and SF concentrations are independently associated with iron depletion at 6 weeks of age [32]. A study conducted by de Waal et al. [30] showed a decrease of iron available for erythropoiesis, when birth weight had already doubled at 6 weeks, and a subsequent increase of ZnPP/H levels. However, these results should be interpreted with caution and take into account the weak study design and subsequent high Risk of Bias (RoB). With regard to iron supplementation, mean birth weight was lower in infants who received supplementation as compared with those who did not [33]. On the other hand, weights of infants at 2 months were not different between the intervention and control groups [24].

In addition, a lower birth weight was found in infants born with iron overload compared to those with normal iron status and latent ID status (serum ferritin <76 ng/mL or <170.7 pmol/L) [36]. Other studies did not find differences between iron supplemented groups and controls in mean weight, length, head circumference, or knee heel length at 6 months [26].

3.5. Iron Deficiency

ID was evaluated despite the different iron status indices used in the studies consulted. Most of the studies revealed that preterm or late preterm neonates usually featured a high risk of ID.

In this sense, indicators of iron status, including Hb levels and the prevalence rates of ID and IDA at 6 months, differed significantly in a dose-dependent manner between the intervention groups. Berglund et al. [26] found that healthy marginally low birth weight infants, both preterm and term, had high risks of developing ID (36%) and IDA (9.9%) at 6 months, especially if they were exclusively breastfed at 6 weeks (ID, 56%; IDA, 18%). This finding is in agreement with the study developed by Uijterschout et al. [33], which reported that ID was present in 18.9% and 4.9% of preterm infants at the age of 4 and 6 months, respectively. IDA was present in 7.7% at the age of 4 months and in 1.4% of preterm infants at the age 6 months. Akkermans et al. [32] showed that iron depletion and iron-depleted anemia in preterm infants at 6 weeks were present at 38.2% and 30.9%, respectively. They showed that preterm infants with a birth weight <1830 g and a SF concentration of <155 μg/L in the first week had a 26.4 times higher risk to develop iron depletion compared to infants with normal birth weights. Similarly, Mukhopadhyay et al. [35] developed a study comparing preterm infants who were small for gestational age, preterm infants appropriate for gestation, and term infants appropriate for gestation. These studies showed that iron stores were significantly decreased in small, preterm, gestational age infants compared to preterm infants appropriate for gestation at birth and at 4 weeks. Likewise, term infants appropriate for gestation had higher iron stores compared to preterm infants appropriate for gestation. In their study, the extent of infants with SF ≤35 μg/L was significantly higher in preterm infants who were small for their gestational age compared to the preterm appropriate for gestation group ($p = 0.01$). Moreover, ferritin levels were significantly lower ($p = 0.006$) those in the preterm appropriate for gestation group compared to the term appropriate for gestation infants. Serum iron also decreased in the preterm small for gestational age group compared to the preterm appropriate for gestation group ($p = 0.002$). Amin et al. [39] showed that 44% of preterm infants had ID (cord ferritin less than 75 ng/mL) and 56% had normal iron status (cord ferritin higher than 75 ng/mL). Once more, these results could be lack of strength to account for the low score obtained in the TREND List and subsequently high RoB.

In this sense, Yamada and Leone [34], indicated that the incidence of anemia was 66.67% in late-preterm newborns at a gestational age of one month post-term and 41.18% in term newborns at one month of age ($p = 0.02$). Hb and Ht decreased mainly in the late-preterm newborns ($p < 0.001$). At birth, differences between the two groups were observed only for Ht ($p = 0.010$), whereas at two months, more dramatic differences in both Hb ($p = 0.020$) and Ht ($p < 0.001$) were reported. Iron concentrations in the late-preterm infants were lower at term ($p = 0.0034$) and at one month post term ($p < 0.001$). TIBC showed lower levels when measured in the late-preterm infants at birth ($p < 0.001$). Transferrin saturation presented a similar pattern and was lower in late-preterm newborns at birth ($p < 0.001$). In a study conducted by de Waal et al. [30] with premature infants at 4 months of age, ID was present in 21.3% and IDA in 8.5% of cases. ZnPP/H levels were used as a possible indicator of iron stores, and at 4 months of age, infants with IDA showed higher levels compared with normal infants ($p < 0.001$). At the same time point, infant with ID did not show any differences in ZnPP/H levels compared with normal infants $p = 0.223$). In addition, a study conducted by Amin et al. [36] revealed that of 131 premature infants, 23% had a latent ID status, 58% had normal iron status while 19% had iron overload status.

Conversely, others studies reported that there were no significant differences between preterm or term infants with regards to ID. Therefore, Molloy et al. [38] observed that all the iron indices examined in preterm infants were elevated above standard normal values in these infants. Sankar et al. [24] described a study with very low birth weight (VLBW) preterm infants, and they showed that the SF at 60 days was not different between VLBW and control groups (full term infants) as well as the mean Ht or Hb at 60 days that were not different for the two groups. Kitajima et al. [37] observed that preterm and full-term infants became anemic during the neonatal period ($p = 0.005$, $p = 0.018$, respectively). At the birth of preterm infants, SF levels were lower than those of full-term infants ($p = 0.011$), and the transferrin saturation of preterm infants did not differ between birth and 1 month after birth levels. Preterm infants showed a significant increase in prohepcidin levels ($p = 0.011$), a finding not shown in full-term infants.

3.6. Iron Supplementation

Iron supplementation was established in 9 studies. Most of the studies supplied iron as elemental iron in ferrous sulfate chemical form, colloidal iron and ferrous succinate mixture while 3 studies did not specify the chemical form of iron supplement [31,34,38]. The differences in iron dosage used in the studies ranged from 1 to 4 mg/kg/day. In addition, the supplementation period started after one or two months of age, but in the majority of the studies, the duration of the supplementation period was not reported. Only four studies specified exactly this period, from 6 weeks to 6 months age [26–29].

3.7. Effects of Iron Supplementation

In the scientific literature, controversies abound on the necessity of t iron supplementation. Some studies showed a positive effect of iron supplementation and its role in preventing or treating ID. In a study with marginally low birth weight preterm or term infants, Berglund et al. [26] showed that iron supplements at a dose of 2 mg/kg per day from 6 weeks to 6 months reduced the risk of IDA effectively, without adverse effects on morbidity or growth observed during the period of supplementation. Taking into account the high score obtained in the Delphi list, and the subsequently low RoB, this dosage and period of supplementation should be considered more plausible than the results reported by other studies. Akkermans et al. [32] revealed that early individualized iron supplementation should be considered as a therapy for high-risk infants with a birth weight of <1830 g and an SF in the first week of <155 μg/L, because late preterm infants have a 26.4 times higher risk of developing iron depletion. In this way, Yamada and Leone [34] suggested the necessity of specific iron supplementation in breastfed late preterm newborns at one month post-term, because greater reductions in Hb/Ht and lower iron stores were found in the preterm infants compared with term newborns.

In contrast, other studies did not show any effect of iron supplementation. Saha et al. [31] suggested that late preterm and term infants (small for gestational age) have adequate iron stores at birth and at 2 months of age; therefore, they do not need iron supplementation until at least the second month of life. In this way, Sankar et al. [24] did not find any difference in SF at 2 months of age between the iron supplementation and control groups when 3 or 4 mg/kg/day (during 2 weeks) of iron supplementation were supplied at 2 weeks of life. In addition, Molloy et al. [38] found elevated iron indices in premature infants prior to establishing iron supplementation. In particular, male premature infants were more susceptible to increased iron levels despite similar numbers of blood transfusions, gestational age, birth weight, albumin, and direct bilirubin. Therefore, iron supplementation should be avoided to prevent iron overload. On the other hand, Amin et al. [36] showed that enteral iron supplementation is unlikely to cause iron overload in premature infants due to the fact that all subjects received enteral iron supplementation of ~2 mg/kg/day, and there was no difference in post-menstrual age when infants reached full volume of enteral feeding (or when iron supplementation was initiated).

Uijterschout et al. [33] proposed that individualized iron supplementation as an alternative to established supplementation therapies be further investigated, and ferritin should be used as a measurement to individualize iron supplementation to late preterm infants. However, it needs to be taken into account that the measurement of ferritin as a sole biomarker may not be appropriate because ferritin is an acute phase protein; therefore, it will not be accurately reflected in the state of inflammation.

3.8. Iron Status and Cognitive Performance

Of the studies included in this systematic review, three directly investigated the effects of iron status on cognitive performance in premature infants. The techniques used to assess cognitive performance were noninvasive.

Berglund et al. [27] tested the auditory brainstem response (ABR) system, which is an auditory evoked potential extracted from ongoing electrical activity in the brain and recorded via electrodes placed on the scalp. This is a neurophysiological method for assessing the development of the central nervous system (CNS) in infants of a marginally low birth weight. They did not report significant differences in absolute ABR latencies between groups, even though the iron supplementation resulted in a significant difference in the prevalence of ID and IDA. This result is surprising taking into account that prolonged latencies were caused by delayed CNS myelination in infants with early IDA. Therefore, they did not find positive or adverse effects of iron supplementation (1–2 mg/kg/day of iron supplementation), on absolute wave I or V latencies at 6 months of age in marginally low birth weight infants, even though they are at risk for ID and IDA. Another study described by Amin et al. [39], revealed that infants with normal iron status had a higher ABR response compared to the ID group ($p = 0.08$).

Another study by Berglund et al. [28], used the Wechsler Preschool and Primary Scale of Intelligence (WPPSI-III) to evaluate cognitive functions in control and premature infants given iron supplementation (1 or 2 mg/kg/day) or placebo (0 mg/kg/day). This study aimed to evaluate the cognitive function and revealed no significant differences between the groups after 3.5 year of the intervention. In addition, in this study, parents were asked to complete the Child Behavior Checklist (CBCL) questionnaire (a widely used caregiver report form that identifies problem behaviors in children and provides a meter stick for measuring whether amounts of behavioral problems have changed over time or across societies and is a complement to other approaches for looking at rates of mental-health issues), before the 3.5-year visit at the study center, in order to assess various types of behavioral and emotional problems. The total CBCL scores revealed that there was a significantly higher prevalence of children above the US subclinical cutoff and above the 90th percentile of the Swedish reference group (Swedish cutoff) in the group who did not receive iron supplementation compared with control subjects and compared with the iron-supplemented groups (1 or 2 mg/kg/day). There was a significant increase in the iron supplemented groups compared to placebo in the CBCL "emotionally reactive" ($p = 0.040$) and "attention problems" ($p = 0.022$) subscales and a similar trend in all subscales except "withdrawn." This study was reported to be the first showing a positive effect of early iron supplementation on neurobehavioral development beyond 2 years of age. This is another surprising result because ID impairs brain development and interferes with neurotransmitter function and biosynthesis, particularly with regard to dopamine and other monoamines.

Berglund et al. [29] tested a psychometric IQ on children 7 years of age using the validated WISC-IV, performed by an experienced authorized pediatric psychologist. Moreover, in this study they also developed a CBCL questionnaire and FTF test (the intervention was performed on LBW infants). These tests assessed behavioral, cognitive, and emotional performance and revealed no significant differences between ID and non-ID groups.

In general the results observed in the different studies of Berglund [26–29] have stronger study designs than the other reports assessing iron status and cognitive performance due to the high scores obtained in the Delphi list, and the subsequently low RoB.

Gupta et al. [25] evaluated neurodevelopment (by a single certified clinical psychologist using the developmental assessment scale) in premature infants who received complementary feeding at 4 months or continued milk feeding and initiated complementary feeding at 6 months (supplemented with iron at 2-3 mg/kg per day). The motor and mental development quotients in the two groups were similar and did not show significant differences between the two groups.

4. Discussion

This review investigated the literature available about the iron status of premature infants and their growth and development. Moreover, it evaluated the effect of iron supplementation in this age group. Premature infants revealed a lower weight, length, and head circumference compared to term infants. The preterm infants with iron depletion had a lower birth weight than those without iron depletion. Iron depletion at 6 weeks of age was independently associated with lower birth weights together with lower SF concentrations [32]. Low iron stores are significantly influenced by low birth weights in preterm infants in general and are more specific in premature infants born after <32 weeks of gestation [40,41]. An impaired placental function caused by pregnancy-induced hypertension, maternal smoking, or gestational diabetes could result in low birth weight [40]. The placenta regulates maternal-foetal iron transport during gestation to supply this essential nutrient to the foetus. Akkermans et al. [32] stated that the most important independent risk factor for iron depletion and iron-depleted anemia in late preterm infants at the age of 6 weeks is a lower birth weight, because they did not find an association between the impaired placental function and depleted iron stores adjusted for birth weight.

4.1. Iron Status

Regarding iron status, results from the studies analyzed showed the absence of consensus on how to define iron stores in preterm infants. SF measurement is the most widely used index to evaluate iron status and it was measured in all studies. Even so, SF is not considered the best iron status parameter by all the authors, because it changes over time and increases during infection and inflammation [30,33,34,42,43]. Therefore, additional biomarkers such as ZnPP/H ratios, which are an indicator of iron available for erythropoiesis, have been proposed [44]. ZnPP/H levels should increase after birth, representing higher iron needs due to rapid growth and iatrogenic iron losses, but should decrease in the first 6 weeks postnatal and return to baseline levels after this period. In premature infants with ID, ferritin levels were low, but ZnPP/H levels were high, showing no association between both parameters. There is a difference between the iron parameters referred to by ferritin and ZnPP/H [30]. Stored iron in the body is reflected by ferritin levels, whereas iron available for erythropoiesis is directly related to the ZnPP/H ratio [30]. In addition to the SF assessment, other traditional iron status parameters such as Hb, Ht, serum iron, transferrin, transferrin saturation, transferrin receptor, and TIBC were evaluated. Most of the data revealed that preterm or late preterm neonates usually showed a lower iron status than term neonates. Yamada and Leone [34] revealed that Hb and Ht significantly reduced while reticulocytes increased during all assessments in the late preterm infants, which might indicate the presence of intense bone marrow stimulation for erythropoiesis. After birth, ferritin levels decreased until two months of age in term and preterm infants [34]. Between 25% and 85% of preterm infants develop evidence of iron deficiency during infancy [34]. Unlike full term infants, in whom the condition typically occurs during the second half of infancy, preterm infants are at risk of developing iron deficiency during their first postnatal months [45]. Iron deficiency is more common in preterm infants from developing countries and in those consuming human milk exclusively without supplementation [46]. A number of factors combine to predispose the premature infants to a negative iron balance. Iron is mostly accumulated during the third trimester of gestation. Total body iron and Hb contents as well as serum and storage iron concentrations are lower in preterm infants. Conditions such as severe maternal iron deficiency, intrauterine growth restriction [47], and chronic blood loss during gestation can further compromise fetal iron endowment.

Mukhopadhyay et al. [35] used ferritin, serum iron, and TIBC as a combination of various biomarkers to assess the iron stores, and they found that preterm infants who were small for their gestational age presented significantly fewer iron stores at birth and at 4 weeks of age compared to preterm infants appropriate for gestation. Low SF levels could be explained by the loss of iron from the storage pool due to chronic hypoxia and increased red cell mass, together with a smaller size of the placenta, which implies lesser surface area, or due to placental vascular disease reducing the transport.

Iron has a crucial role in many metabolic pathways, especially in erythropoiesis and neurodevelopment [1]. Specifically, Hb formation, which is crucial for oxygen transport to tissues, is dependent on iron absorbed from the diet and storage. If the absorption of dietary iron is insufficient or the deposits are depleted, Hb still can be synthesized until serum iron decreases. After birth, higher environmental oxygen concentrations, a decrease in fetal Hb, an increase in adult-type Hb, polycythemic conditions, and Hb oxygen saturation promote great tissue oxygenation, thereby decreasing the stimulation of erythropoietin production and reducing erythrocyte release. Therefore, Hb levels are reduced by 30% to 50%, reaching their minimum value at six to twelve weeks after birth and one to four weeks earlier in preterm infants [48]. Furthermore, the fetal iron storage of preterm infants is incomplete before their birth, increasing the risk of ID and even IDA during the early postnatal period [41,49,50].These results suggest that preterm infants have increased iron needs after birth [34]. In addition, irreversible processes such as impaired neurological development in preterm infants may be caused by ID, even without anemia [51].

Although, some studies did not find lower iron parameters in premature infants after birth, Molloy et al. [38] found that in preterm infants some iron indices were elevated above the standard normal values in these infants. In this way, Saha et al. [31] observed that the Hb and SF of all late preterm infants were within a normal range at 2 months of age, which implies adequate stores at birth to meet requirements. The inflammation/infection process could explain these levels, which has been associated with an increase of SF.

Taking into account iron status in preterm infants, it seems highly relevant to study the presence or absence of ID or IDA in this population. Premature infants have increased risk to develop ID, which is the most common micronutrient deficiency in the world [52], and also IDA, due to low endowment of iron stores, long periods of parenteral nutrition without iron supplementation, phlebotomy loss, and rapid catch-up growth and development [4]. Most of the selected studies revealed that ID and IDA were present in premature infants, being more common when the premature infants were exclusively breastfed [26,28,53]. Moreover, ID and IDA have been associated with long lasting abnormal neurodevelopmental outcomes in full-term, normal weight infants, during the critical period of brain development [1,54], due to their effect on cognitive and psychomotor development, even after iron treatment [55]. In addition, growth and functioning of other organ systems such as skeletal muscle, the heart, and the gastrointestinal tract could be also affected by ID [8,10].

Dewey et al. [56] reported that iron supplementation (1 mg/(kg/day)) to breastfed infants may benefit infants with low hemoglobin concentrations but may present adverse effects, such as deficits in gains in length and head circumference and an increased risk of diarrhea for those with normal hemoglobin concentrations. Majumdar et al. [57] similarly found that iron supplementation (2 ng/(kg/day)) benefited iron-deplete children aged 6–24 months but led to deficits in weight and linear weight gain in those who were iron replete.

4.2. Iron Supplementation

According to these results and taking into account the importance of avoiding ID in this population, iron supplementation of preterm neonates seems vital because rapidly depleting inadequate iron stores cannot meet the iron requirement of enhanced erythropoiesis during this period. However, previous studies on iron supplementation in preterm infants have yielded conflicting results [45,58]. Although iron is essential for children's development, iron overload should be considered because of the potential risk of iron excess and the poorly developed antioxidant mechanisms in preterm infants, facts that reinforce the idea of avoiding indiscriminate iron supplementation in this population [4,59]. Iron overload mediates oxidative stress and may also contribute to free oxygen radical injuries that are typical of premature infants, such as chronic lung disease and retinopathy from prematurity [60,61]. IDA was reduced effectively with a dose of 2 mg/kg per day from 6 weeks to 6 months [26], and non-adverse effects on morbidity or growth were found. In addition, as most pathogenic organisms need iron as an essential nutrient, it has been suggested that iron supplementation may increase the risk of infections, but in this study, the supplementation did not affect infections or other pathological symptoms. Moreover, another study suggests that iron supplementation is justified in late preterm infants with a birth weight of <1830 g and ferritin concentrations of <155 µg/L in the first week [32].

In contrast, Sankar et al. [24] did not find any association with iron supplementation at 2 weeks of life, improving neither SF nor hematological parameters at 2 months of age at preterm. This could possibly be due to inadequate iron absorption, inappropriate dosage or a "contamination" effect (iron preparation used in this study contains colloidal iron in the form of non-ionic, microfine particles). By the second month of life, erythropoiesis is increased and iron stores are rapidly utilized. Although SF levels decreased after birth due to the utilization of iron stores, both the SF and Hb of all premature infants were within normal range at 2 months of age [31]. Therefore, iron stores at birth were adequate to meet the requirements of these infants for at least 2 months of life, and iron supplementation is not necessary during this period. In this sense, Molloy et al. [38] found elevated iron indices in infants prior to commencing iron supplements, so an evaluation of iron stores after birth is essential to prevent potential iron overload and unnecessary iron supplementation. However, another explanation could be unmeasured confounding factors in these studies.

4.3. *Cognitive Performance*

Regarding the effect of iron supplementation on cognitive performance, Berglund et al. [27] evaluated the effect of 1–2 mg/kg/day of iron supplementation on ABR as a noninvasive, objective, neurophysiologic method for assessing the development of the central nervous system. ABR latencies measure the conduction speed in the auditory system from the cochlea to the inferior colliculus in the upper brainstem. From late gestation through the first 3–6 months of life, these latencies experience a rapid decrease continuing at a slower rate during the first 3–5 years of life, indicator of CNS myelination during infancy for preterm and term infants. In this study, no significant positive or adverse effects in absolute ABR latencies between 0 mg/kg/day (placebo), 1 mg/kg/day, or 2 mg/kg/day groups were found at 6 months of age, even though the iron supplementation resulted in a significant decrease in the prevalence of ID and IDA. Several possible explanations could be proposed. The duration or the severity were not at a magnitude high enough to result in prolonged ABR latencies, or these infants did not have a vulnerable phase of ABR maturation when ID occurred. Finally, ABR latencies could be affected later as a delayed effect of ID. Therefore, these results could suggest that ABR is not a sensitive indicator of impaired neurological development in MLBW infants or that severe ID does not cause such impairment at 6 months of age. In addition, a cord ferritin concentration is associated with ABR in premature infants. Cord ferritin lower than 75 ng/mL is associated with poor performance in motor skills and language development [62]. Iron status influences auditory neural maturation and infants with ID have a normal auditory neural maturation compared with a normal iron status [39]

Berglund et al. [28] did not find any effect of iron supplementation on cognition at 3.5 years of age. Furthermore, marginally premature infants with low birth weight did not present a higher risk of cognitive problems compared with control infants, although the cognitive test performed was not accurate enough to predict later cognitive disabilities. Conversely, non-supplemented children have a significantly increased prevalence of behavioral problems; therefore, preventive iron supplementation is related to improved neurobehavioral development in infants at risk of ID. An iron intake of >1 mg/kg per day could be sufficient to decrease the risk of behavioral problems due to a correlation between iron intake and CBCL. Therefore, this study showed that iron supplementation reduced the risk of behavioral problems to a level similar to the controls. Many animal studies have reported that ID impairs brain development [28]. In addition, Berglund et al. [29] reported that this protective effect, which includes aggressive and rule-breaking behavior, persists until 7 years of age in children.

Iron plays a crucial role in myelination, dendritic growth, and synaptogenesis. Therefore, ID could interfere with neurotransmitter function, mainly with regard to dopamine and other monoamines, which interestingly are closely associated with behavior [54]. Nonetheless, Berglund et al. [28] did not find a significant association between the behavioral outcome and iron status at 12 weeks or 6 months in infants. A possible explanation could be the lack of consensus biomarkers to define ID in infancy or because the used iron indicators lack specificity and sensitivity. Moreover, the iron availability in the CNS is not necessarily reflected by the biomarkers of iron status that reflect the iron status of peripheral blood and probably that of the bone marrow and the liver. In addition, Berglund et al. [29] revealed that the iron status at 6 months did not correlate with behavioral outcomes at either 3.5 or 7 years in LBW infants. This suggests that less iron is available for the brain tissue before ID can be detected in the blood by conventional haematological indices, since iron requirements for erythropoiesis are b prioritized over the brain iron supply [63].

Finally, Gupta et al. [25] evaluated neurodevelopment using a Developmental Assessment Scale for Indian Infants (DASII), a validated Indian adaptation of Bayley-II [64], in preterm infants who received iron supplementation as a standard. They did not find differences between infants who initiated complementary feeding at 4 months and those who continued with breast milk feeding and initiated complementary feeding at 6 months. The infants showed similar motor and mental development quotients. This study revealed that despite the fact that the infants received iron supplementation with either food or milk, their iron stores were greatly depleted.

5. Conclusions

In summary, the findings of this systematic review provide evidence that premature infants require iron supplementation in most of the cases and are susceptible to the development of ID or IDA, which could affect postnatal cognitive development and behavior. However, although ID impairs brain development due to its essential role in myelination, dendritic growth, synaptogenesis, and neurotransmitter function, motor, mental, and development parameters were not affected by iron supplementation in preterm neonates. Infants with iron depletion and overload showed a lower birth weight than those with a normal iron status. Taking into account these considerations, future research with more robust experimental designs and lower RoB are needed to achieve more solid conclusions.

Author Contributions: J.D.-C. and J.M.-F. designed the study; J.M.-F. and J.D.-C. performed the literature search; J.M.-F. and J.D.-C. screened the articles; J.J.O validated the screened articles; J.M.-F. and J.D.-C. wrote original review; G.O.L.-D. and J.J.O reviewed and edited the manuscript. All authors interpreted the data, wrote the paper, and have read and approved the final manuscript.

Funding: This research received no external funding.

Acknowledgments: Jorge Moreno-Fernandez was supported by fellowship from the Ministry of Education, Culture and Sport (Spain) and is grateful to the Excellence Ph.D. Program "Nutrición y Ciencias de los Alimentos" from the University of Granada. The authors are grateful to Susan Stevenson for her efficient support in the revision with the English language.

Conflicts of Interest: The authors declare no conflict of interest.

References

1. Lozoff, B.; Beard, J.; Connor, J.; Barbara, F.; Georgieff, M.; Schallert, T. Long-lasting neural and behavioral effects of iron deficiency in infancy. *Nutr. Rev.* **2006**, *64*, S34–S43. [CrossRef] [PubMed]
2. McLean, E.; Cogswell, M.; Egli, I.; Wojdyla, D.; de Benoist, B. Worldwide prevalence of anaemia, who vitamin and mineral nutrition information system, 1993–2005. *Public Health Nutr.* **2009**, *12*, 444–454. [CrossRef] [PubMed]
3. Collard, K.J. Iron homeostasis in the neonate. *Pediatrics* **2009**, *123*, 1208–1216. [CrossRef] [PubMed]
4. Rao, R.; Georgieff, M.K. Iron therapy for preterm infants. *Clin. Perinatol.* **2009**, *36*, 27–42. [CrossRef]
5. Lorenz, L.; Peter, A.; Poets, C.F.; Franz, A.R. A review of cord blood concentrations of iron status parameters to define reference ranges for preterm infants. *Neonatology* **2013**, *104*, 194–202. [CrossRef]
6. Wang, M. Iron deficiency and other types of anemia in infants and children. *Am. Fam. Phys.* **2016**, *93*, 270–278.
7. Blayney, L.; Bailey-Wood, R.; Jacobs, A.; Henderson, A.; Muir, J. The effects of iron deficiency on the respiratory function and cytochrome content of rat heart mitochondria. *Circ. Res.* **1976**, *39*, 744–748. [CrossRef]

8. Mackler, B.; Grace, R.; Finch, C.A. Iron deficiency in the rat: Effects on oxidative metabolism in distinct types of skeletal muscle. *Pediatr. Res.* **1984**, *18*, 499–500. [CrossRef] [PubMed]
9. Ercan, O.; Ulukutlu, L.; Ozbay, G.; Arda, O. Intestinal effects of iron deficiency anemia in children. *Turkish J. Pediatr.* **1991**, *33*, 85–98.
10. Berant, M.; Khourie, M.; Menzies, I.S. Effect of iron deficiency on small intestinal permeability in infants and young children. *J. Pediatr. Gastroenterol. Nutr.* **1992**, *14*, 17–20. [CrossRef]
11. Aggett, P.J. Trace elements of the micropremie. *Clin. Perinatol.* **2000**, *27*, 119–129. [CrossRef]
12. Beard, J.L. Why iron deficiency is important in infant development. *J. Nutr.* **2008**, *138*, 2534–2536. [CrossRef]
13. Kon, N.; Tanaka, K.; Sekigawa, M.; Negishi, Y.; Yoshikawa, N.; Hisata, K.; Shoji, H.; Shimizu, T. Association between iron status and neurodevelopmental outcomes among vlbw infants. *Brain Dev.* **2010**, *32*, 849–854. [CrossRef]
14. Grantham-McGregor, S.; Ani, C. A review of studies on the effect of iron deficiency on cognitive development in children. *J. Nutr.* **2001**, *131*, 649S–666S; discussion 666S–668S. [CrossRef] [PubMed]
15. Angulo-Kinzler, R.M.; Peirano, P.; Lin, E.; Garrido, M.; Lozoff, B. Spontaneous motor activity in human infants with iron-deficiency anemia. *Early Human Dev.* **2002**, *66*, 67–79. [CrossRef]
16. Gera, T.; Sachdev, H.P.; Nestel, P.; Sachdev, S.S. Effect of iron supplementation on haemoglobin response in children: Systematic review of randomised controlled trials. *J. Pediatr. Gastroenterol. Nutr.* **2007**, *44*, 468–486. [CrossRef]
17. Sachdev, H.; Gera, T.; Nestel, P. Effect of iron supplementation on mental and motor development in children: Systematic review of randomised controlled trials. *Public Health Nutr.* **2005**, *8*, 117–132. [CrossRef]
18. Sachdev, H.P.S.; Gera, T.; Nestel, P. Effect of iron supplementation on physical growth in children: Systematic review of randomised controlled trials. *Public Health Nutr.* **2007**, *9*, 904–920. [CrossRef]
19. Luckmann, R. Evidence-based medicine: How to practice and teach ebm, 2nd edition: By david l. Sackett, sharon e. Straus, w. Scott richardson, william rosenberg, and r. Brian haynes, churchill livingstone, 2000. *J. Intensive Care Med.* **2001**, *16*, 155–156. [CrossRef]
20. Villar Álvarez, F.; Estrada Lorenzo, J.M.; Pérez Andrés, C.; Rebollo Rodríguez, M.J. Estudio bibliométrico de los artículos originales de la revista española de salud pública (1991–2000): Parte tercera: Análisis de las referencias bibliográficas. *Rev. Esp. Salud Publica* **2007**, *81*, 247–259. [CrossRef] [PubMed]
21. Wanden-Berghe, C.; Sanz-Valero, J. Systematic reviews in nutrition: Standardized methodology. *Br. J. Nutr.* **2012**, *107* (Suppl. 2), S3–S7. [CrossRef]
22. Verhagen, A.P.; de Vet, H.C.; de Bie, R.A.; Kessels, A.G.; Boers, M.; Bouter, L.M.; Knipschild, P.G. The delphi list: A criteria list for quality assessment of randomized clinical trials for conducting systematic reviews developed by delphi consensus. *J. Clin. Epidemiol.* **1998**, *51*, 1235–1241. [CrossRef]
23. Des Jarlais, D.C.; Lyles, C.; Crepaz, N. Improving the reporting quality of nonrandomized evaluations of behavioral and public health interventions: The trend statement. *Am. J. Public Health* **2004**, *94*, 361–366. [CrossRef] [PubMed]
24. Sankar, M.J.; Saxena, R.; Mani, K.; Agarwal, R.; Deorari, A.K.; Paul, V.K. Early iron supplementation in very low birth weight infants–a randomized controlled trial. *Acta Paediatr. (Oslo, Norway)* **2009**, *98*, 953–958. [CrossRef]
25. Gupta, S.; Agarwal, R.; Aggarwal, K.C.; Chellani, H.; Duggal, A.; Arya, S.; Bhatia, S.; Sankar, M.J.; Sreenivas, V.; Jain, V.; et al. Complementary feeding at 4 versus 6 months of age for preterm infants born at less than 34 weeks of gestation: A randomised, open-label, multicentre trial. *Lancet Global Health* **2017**, *5*, e501–e511. [CrossRef]
26. Berglund, S.; Westrup, B.; Domellof, M. Iron supplements reduce the risk of iron deficiency anemia in marginally low birth weight infants. *Pediatrics* **2010**, *126*, e874–e883. [CrossRef]
27. Berglund, S.K.; Westrup, B.; Haraldsson, E.; Engstrom, B.; Domellof, M. Effects of iron supplementation on auditory brainstem response in marginally lbw infants. *Pediatr. Res.* **2011**, *70*, 601–606. [CrossRef]
28. Berglund, S.K.; Westrup, B.; Hagglof, B.; Hernell, O.; Domellof, M. Effects of iron supplementation of lbw infants on cognition and behavior at 3 years. *Pediatrics* **2013**, *131*, 47–55. [CrossRef]
29. Berglund, S.K.; Chmielewska, A.; Starnberg, J.; Westrup, B.; Hagglof, B.; Norman, M.; Domellof, M. Effects of iron supplementation of low-birth-weight infants on cognition and behavior at 7 years: A randomized controlled trial. *Pediatr. Res.* **2018**, *83*, 111–118. [CrossRef] [PubMed]

30. de Waal, C.G.; Uijterschout, L.; Abbink, M.; Boersma, B.; Vos, P.; Rovekamp, W.W.; Hudig, F.; Akkermans, M.D.; van Goudoever, J.B.; Brus, F. Zinc protoporphyrin/heme ratio as parameter of iron status in moderately preterm infants: Natural course and associations in the first 4 months. *J. Perinatol.* **2017**, *37*, 690–694. [CrossRef] [PubMed]
31. Saha, B.; Jeeva Sankar, M.; Gupta, S.; Agarwal, R.; Gupta, N.; Deorari, A.; Paul, V.K. Iron stores in term and late preterm small for gestational age and appropriate for gestational age neonates at birth and in early infancy. *Indian J. Pediatr.* **2016**, *83*, 622–627. [CrossRef]
32. Akkermans, M.D.; Uijterschout, L.; Abbink, M.; Vos, P.; Rovekamp-Abels, L.; Boersma, B.; van Goudoever, J.B.; Brus, F. Predictive factors of iron depletion in late preterm infants at the postnatal age of 6 weeks. *Eur. J. Clin. Nutr.* **2016**, *70*, 941–946. [CrossRef] [PubMed]
33. Uijterschout, L.; Domellof, M.; Abbink, M.; Berglund, S.K.; van Veen, I.; Vos, P.; Rovekamp, L.; Boersma, B.; Hudig, C.; Vos, R.; et al. Iron deficiency in the first 6 months of age in infants born between 32 and 37 weeks of gestational age. *Eur. J. Clin. Nutr.* **2015**, *69*, 598–602. [CrossRef] [PubMed]
34. Yamada, R.T.; Leone, C.R. Hematological and iron content evolution in exclusively breastfed late-preterm newborns. *Clinics (Sao Paulo, Brazil)* **2014**, *69*, 792–798. [CrossRef]
35. Mukhopadhyay, K.; Yadav, R.K.; Kishore, S.S.; Garewal, G.; Jain, V.; Narang, A. Iron status at birth and at 4 weeks in preterm-sga infants in comparison with preterm and term-aga infants. *J. Matern. Fetal. Neonatal. Med.* **2012**, *25*, 1474–1478. [CrossRef]
36. Amin, S.B.; Scholer, L.; Srivastava, M. Pre-discharge iron status and its determinants in premature infants. *J. Matern. Fetal. Neonatal. Med.* **2012**, *25*, 2265–2269. [CrossRef]
37. Kitajima, J.; Ohga, S.; Kinjo, T.; Ochiai, M.; Takahata, Y.; Honjo, S.; Hara, T. Serum prohepcidin concentrations at birth and 1 month after birth in premature infants. *Pediatr. Blood Cancer* **2011**, *56*, 267–272. [CrossRef]
38. Molloy, E.J.; El-Khuffash, A.; Bieda, A.; Jelinek, M.M.; Baley, J. Elevated iron indices in preterm infants: Association with male gender. *Am. J. Perinatol.* **2009**, *26*, 7–11. [CrossRef]
39. Amin, S.B.; Orlando, M.; Eddins, A.; MacDonald, M.; Monczynski, C.; Wang, H. In utero iron status and auditory neural maturation in premature infants as evaluated by auditory brainstem response. *J. Pediatr.* **2010**, *156*, 377–381. [CrossRef]
40. Siddappa, A.M.; Rao, R.; Long, J.D.; Widness, J.A.; Georgieff, M.K. The assessment of newborn iron stores at birth: A review of the literature and standards for ferritin concentrations. *Neonatology* **2007**, *92*, 73–82. [CrossRef]
41. Domellof, M.; Braegger, C.; Campoy, C.; Colomb, V.; Decsi, T.; Fewtrell, M.; Hojsak, I.; Mihatsch, W.; Molgaard, C.; Shamir, R.; et al. Iron requirements of infants and toddlers. *J. Pediatr. Gastroenterol. Nutr.* **2014**, *58*, 119–129. [CrossRef]
42. Tomkins, A. Assessing micronutrient status in the presence of inflammation. *J. Nutr.* **2003**, *133*, 1649s–1655s. [CrossRef] [PubMed]
43. Jain, S.; Narayan, S.; Chandra, J.; Sharma, S.; Jain, S.; Malhan, P. Evaluation of serum transferrin receptor and stfr ferritin indices in diagnosing and differentiating iron deficiency anemia from anemia of chronic disease. *Indian J. Pediatr.* **2010**, *77*, 179–183. [CrossRef]
44. Metzgeroth, G.; Adelberger, V.; Dorn-Beineke, A.; Kuhn, C.; Schatz, M.; Maywald, O.; Bertsch, T.; Wisser, H.; Hehlmann, R.; Hastka, J. Soluble transferrin receptor and zinc protoporphyrin–competitors or efficient partners? *Eur. J. Haematol.* **2005**, *75*, 309–317. [CrossRef] [PubMed]
45. Franz, A.R.; Mihatsch, W.A.; Sander, S.; Kron, M.; Pohlandt, F. Prospective randomized trial of early versus late enteral iron supplementation in infants with a birth weight of less than 1301 grams. *Pediatrics* **2000**, *106*, 700–706. [CrossRef]
46. Borigato, E.V.; Martinez, F.E. Iron nutritional status is improved in brazilian preterm infants fed food cooked in iron pots. *J. Nutr.* **1998**, *128*, 855–859. [CrossRef] [PubMed]
47. Olivares, M.; Llaguno, S.; Marin, V.; Hertrampf, E.; Mena, P.; Milad, M. Iron status in low-birth-weight infants, small and appropriate for gestational age. A follow-up study. *Acta Paediatr. (Oslo, Norway)* **1992**, *81*, 824–828. [CrossRef]
48. Rao, R.; Georgieff, M.K. Iron in fetal and neonatal nutrition. *Semin. Fetal Neonatal Med.* **2007**, *12*, 54–63. [CrossRef]
49. Griffin, I.; Cooke, R.J. Iron retention in preterm infants fed low iron intakes: A metabolic balance study. *Early Human Dev.* **2010**, *86* (Suppl. 1), 49–53. [CrossRef]

50. Long, H.; Yi, J.-M.; Hu, P.-L.; Li, Z.-B.; Qiu, W.-Y.; Wang, F.; Zhu, S. Benefits of iron supplementation for low birth weight infants: A systematic review. *BMC pediatr.* **2012**, *12*, 99. [CrossRef] [PubMed]
51. Walker, S.P.; Wachs, T.D.; Grantham-McGregor, S.; Black, M.M.; Nelson, C.A.; Huffman, S.L.; Baker-Henningham, H.; Chang, S.M.; Hamadani, J.D.; Lozoff, B.; et al. Inequality in early childhood: Risk and protective factors for early child development. *Lancet (London, England)* **2011**, *378*, 1325–1338. [CrossRef]
52. World Health Organization (Ed.) *Iron Deficiency Anaemia: Assessment, Prevention and Control: A Guide for Programme Managers*; WHO: Geneva, Switzerland, 2001.
53. Meinzen-Derr, J.K.; Guerrero, M.L.; Altaye, M.; Ortega-Gallegos, H.; Ruiz-Palacios, G.M.; Morrow, A.L. Risk of infant anemia is associated with exclusive breast-feeding and maternal anemia in a mexican cohort. *J. Nutr.* **2006**, *136*, 452–458. [CrossRef] [PubMed]
54. Lozoff, B.; Georgieff, M.K. Iron deficiency and brain development. *Semin. Pediatr. Neurol.* **2006**, *13*, 158–165. [CrossRef] [PubMed]
55. Lozoff, B. Iron deficiency and child development. *Food Nutr. Bull.* **2007**, *28*, S560–S571. [CrossRef] [PubMed]
56. Domellöf, M.; Hernell, O.; Landa Rivera, L.; Lönnerdal, B.; Cohen, R.J.; Dewey, K.G. Iron supplementation affects growth and morbidity of breast-fed infants: Results of a randomized trial in sweden and honduras. *J. Nutr.* **2002**, *132*, 3249–3255.
57. Majumdar, I.; Paul, P.; Talib, V.H.; Ranga, S. The effect of iron therapy on the growth of iron-replete and iron-deplete children. *J. Trop. Pediatr.* **2003**, *49*, 84–88. [CrossRef]
58. Lundstrom, U.; Siimes, M.A.; Dallman, P.R. At what age does iron supplementation become necessary in low-birth-weight infants? *J. Pediatr.* **1977**, *91*, 878–883. [CrossRef]
59. Perrone, S.; Tataranno, L.M.; Stazzoni, G.; Ramenghi, L.; Buonocore, G. Brain susceptibility to oxidative stress in the perinatal period. *J. Matern. Fetal. Neonatal. Med.* **2015**, *28* (Suppl. 1), 2291–2295. [CrossRef]
60. Inder, T.E.; Clemett, R.S.; Austin, N.C.; Graham, P.; Darlow, B.A. High iron status in very low birth weight infants is associated with an increased risk of retinopathy of prematurity. *J. Pediatr.* **1997**, *131*, 541–544. [CrossRef]
61. Cooke, R.W.I.; Drury, J.A.; Yoxall, C.W.; James, C. Blood transfusion and chronic lung disease in preterm infants. *Eur. J. Pediatr.* **1996**, *156*, 47–50. [CrossRef]
62. Tamura, T.; Goldenberg, R.L.; Hou, J.; Johnston, K.E.; Cliver, S.P.; Ramey, S.L.; Nelson, K.G. Cord serum ferritin concentrations and mental and psychomotor development of children at five years of age. *J. Pediatr.* **2002**, *140*, 165–170. [CrossRef] [PubMed]
63. Zamora, T.G.; Guiang, S.F., 3rd; Widness, J.A.; Georgieff, M.K. Iron is prioritized to red blood cells over the brain in phlebotomized anemic newborn lambs. *Pediatr. Res.* **2016**, *79*, 922–928. [CrossRef]
64. Pathak, P. Baroda norms of motor and mental development of indian babies from 1 to 30 months. *Psychol. Stud.* **1993**, *38*, 142–149.

nutrients

MDPI

Article

Maternal and Early Life Iron Intake and Risk of Childhood Type 1 Diabetes: A Danish Case-Cohort Study

Steffen Ullitz Thorsen [1,2,*], Thorhallur I. Halldorsson [2,3], Anne A. Bjerregaard [2], Sjurdur F. Olsen [2] and Jannet Svensson [1,4]

1 Copenhagen Diabetes Research Center (CPH-DIRECT), Department of Paediatrics, Herlev Hospital, University of Copenhagen, Herlev Ringvej 75, 2730 Herlev, Denmark; Jannet.Svensson@regionh.dk

2 Centre for Fetal Programming, Department of Epidemiology Research, Statens Serum Institut, 2300 Copenhagen S, Denmark; LUR@ssi.dk (T.I.H.); ANNE@ssi.dk (A.A.B.); SFO@ssi.dk (S.F.O.)

3 Unit for Nutrition Research, Faculty of Food Science and Nutrition, School of Health Sciences, University of Iceland, 101 Reykjavík, Iceland

4 Department of Clinical Medicine, Faculty of Health and Medical Sciences, University of Copenhagen, Blegdamsvej 3B, 2200 Copenhagen N, Denmark

* Correspondence: s.u.thorsen@gmail.com; Tel.: +45-22288995

Received: 14 February 2019; Accepted: 22 March 2019; Published: 29 March 2019

Abstract: Background: Iron overload has been associated with diabetes. Studies on iron exposure during pregnancy and in early life and risk of childhood type 1 diabetes (T1D) are sparse. We investigated whether iron supplementation during pregnancy and early in life were associated with risk of childhood T1D. Methods: In a case-cohort design, we identified up to 257 children with T1D (prevalence 0.37%) from the Danish National Birth Cohort through linkage with the Danish Childhood Diabetes Register. The primary exposure was maternal pure iron supplementation (yes/no) during pregnancy as reported in interview two at 30 weeks of gestation (n = 68,497 with iron supplement data). We estimated hazard ratios (HRs) using weighted Cox regression adjusting for multiple confounders. We also examined if offspring supplementation during the first 18 months of life was associated with later risk of T1D. Results: Maternal iron supplementation was not associated with later risk of T1D in the offspring HR 1.05 (95% CI: 0.76–1.45). Offspring intake of iron droplets during the first 18 months of life was inversely associated with risk of T1D HR 0.74 (95% CI: 0.55–1.00) (p_{trend} = 0.03). Conclusions: Our large-scale prospective study demonstrated no harmful effects of iron supplementation during pregnancy and in early life in regard to later risk of childhood T1D in the offspring.

Keywords: diabetes mellitus; type 1; iron; pregnancy; fetal programming; infant; newborn

1. Introduction

Type 1 diabetes (T1D) is a chronic immune-mediated disease with a selective destruction of pancreatic β-cells leading to life-long insulin dependency and severe long-term complications [1,2]. T1D is triggered by both genetic and environmental factors [1]. Genetic susceptibility loci have been extensively identified [3], but only a few environmental factors have been proposed to associate with T1D [4]. The incidence of T1D is increasing at an annual rate of 3–5% in Denmark and many other countries [5,6], which suggest that yet unknown environmental factors play an important role in T1D etiopathology.

Iron, the most abundant metal in the human body, is an essential trace element for a plethora of cellular functions e.g., a normal maturation and function of the immune system [7,8]. Though, excess iron may be harmful e.g., due to formation of reactive oxygen species (ROS) by participating in the

Fenton reaction [9]. Pancreatic β-cells possesses a classical iron metabolism [10] and are especially sensitive to ROS due to finite antioxidant defenses [11,12].

Iron deficiency and iron deficiency anemia are common conditions in pregnant women and during early childhood in both the European population and worldwide [13–15]. Iron requirement increases during pregnancy and therefore it is recommended that pregnant women take iron supplementations [13,14], and breastfed children are supplemented. However, in Western countries, such recommendations, on top of an already sufficient intake, could result in excess iron intake, in a subgroup of women. Further, use of several supplements or not following the recommended dosing can easily result in doses above the recommended upper intake levels [16,17]. Noteworthy, a study originated from the Danish National Birth Cohort (DNBC) reported that 37% of Danish pregnant women had an iron supplementation intake above the recommended 50–70 mg per day [16], which could lead to unfavorable iron concentrations in both the mother and child, but data regarding adverse effects of high maternal iron supplementation are still limited [18].

Though, much is still to be learned about the transport of iron from the maternal circulation through the placenta to the fetus, placental iron transport during pregnancy is mainly driven by the fetal need, but maternal iron deficiency or excess may compromise the fetal iron homeostasis [19–21]. It can be hypothesized that such extremes in maternal supply may have adverse consequences for the unborn child [22]. To our knowledge, only one study has examined the association between maternal iron status and intake and later risk of childhood T1D in the offspring. In the prospective Norwegian Mother and Child Cohort Study a positive association was found between maternal iron supplementation and risk of childhood T1D [23]. Further, only five relatively small-sized retrospective studies with moderate to high assessment quality have examined the association between early life iron exposure and later risk of childhood T1D, and results have been mixed [24,25].

Following up on the findings from the large Norwegian study we tested the hypothesis that there is an association between in utero and early life iron exposure and risk of childhood T1D, using pure iron supplementation data from the unique prospective DNBC [26] obtained through telephone interviews at 30 weeks of gestation, six and 18 months postpartum. Further, iron from both pure iron supplements and multivitamins were obtained from a 360-item food frequency questionnaire (FFQ) sent to the women at approximately 25 weeks of gestation [16].

2. Materials and Methods

2.1. Overview of Study Design

The study is a cohort study including maternal-child pairs from the DNBC [26] enrolling pregnant Danish women from 1996 until 2002. Altogether 91,745 mothers were recruited. Some women participated more than once leading to a total number of 101,042 pregnancies. Mothers were recruited at their first visit around 6–10 weeks of gestation at their general practitioner. All pregnant women residents in Denmark who were fluent in Danish were invited. Approximately 35% of all births in Denmark during the recruitment period were involved.

2.2. Study Sample and Identification of T1D

Approximately 260 controls per case exist in the cohort (sampling ratio 1:260). The proportion of any use of pure iron supplements during pregnancy in DNBC is 0.79. With 13,196 out of 63,931 women reporting use of pure iron supplements in pregnancy and the prevalence of offspring T1D being 0.37% we have a power of 80% to detect an absolute difference in risk of 1.6 when comparing users to non-users.

Patients were found through linkage with the Danish Childhood Diabetes Register (DanDiabKids), covering children aged 0–18 years diagnosed with T1D [27]. DanDiabKids was initiated in 1996 and is validated annually against the National Patient Discharge Register. DanDiabKids have nearly complete nationwide coverage and record high-quality prospective data on children with T1D. The time of clinical T1D diagnosis was set as the first day of insulin treatment in accordance with International Society for Pediatric and Adolescent Diabetes (ISPAD) guidelines. We included all children diagnosed with T1D who had available iron supplemental data for either one of the three time periods (pregnancy, and at six and 18 months of follow-up).

2.3. Exposure Assessment

2.3.1. Maternal Iron Supplementation

Following up on the findings form the Norwegian Mother and Child Cohort study [23], our primary exposure was maternal use of pure iron supplements (supplements only containing iron) in early (<20 weeks) and late (≥20 weeks) pregnancy. Supplemental intake for these two time points was based on maternal report in telephone interviews conducted in week 30 of gestation and six months postpartum, respectively. In these interviews, the women were asked "Have you been taking iron pills during pregnancy?". If they answered "yes" they were asked in which weeks of gestation they had been taking supplements, ranging from the beginning of pregnancy until the time of the 30-week interview; and the interview conducted at six months postpartum covered the remaining period of their pregnancy. In addition, we also assessed the total supplemental intake of iron as reported when the women filled out an FFQ in week 25 of gestation. The FFQ covered the supplemental intake of the previous four weeks. The women were asked to report all supplements they had been taken and for each label the amount of micronutrient, e.g., iron, and the daily dose [16].

In our study, the number of subjects available for analyses varied depending on the source of information. Information gained from interviews in week 30 of gestation and six months postpartum regarding use of pure iron supplements during pregnancy ranged from 63,931 to 64,456. Information on use of any iron containing supplements as reported by the mother in week 25 of gestation was available for 68,240.

2.3.2. Infant Iron Supplementation

Iron supplement exposure during early life was extracted from the fourth telephone interview conducted at 18 months postpartum where the mother was asked if her child had received iron droplets (yes/no), and the duration of the supplementation (months). The women were asked "For how many months did he/she receive iron droplets on a regular basis?", and the answered categories were as follows: i. under 1 month; ii. 1–2 months; iii. 3–6 months; and iv. over 6 months (https://www.dnbc.dk/-/media/arkiv/projekt-sites/dnbc/kodeboeger/interviews-1-4/code_book_interview_4.pdf?la=en).

2.3.3. Other Variables

The following *a priori* covariates were included in the main adjusted models based on previous work and the literature: maternal age at delivery (continuous), pre-pregnancy body mass index (BMI) (<18.5, 18.5–24.9, 25–30, and ≥30 kg/m^2), parity (nulliparous (yes/no)), smoking during pregnancy (never, occasional, and yes), parental socioeconomic status (high, medium, low, or student), duration of breastfeeding (0, 0–6, or 6+ months), caesarean section (yes/no), gestational age at delivery (continuous) (Table 1). The following covariates were considered for sensitivity analyses: maternal anaemia (yes/no), maternal celiac disease (yes/no), and maternal T1D (yes/no) (Table 1). Specification of included covariates in the adjusted models is presented in the footnotes of Tables 2 and 3.

2.4. Statistical Approach

All details of the analysis plan were determined *a priori*. We used the mean and standard deviation (SD) to describe normally distributed continuous variables and percentages for dichotomous variables. Cox regression was used for examining associations between maternal or offspring intake of iron-containing supplements and later risk of developing T1D in the offspring. Offspring age from birth up to May 2016 was used as the underlying timescale—start of follow-up was 18 months in the offspring analyses (time of iron droplet data collection). The effect estimates were reported as hazard ratios (HR) with 95% CI. As women could enter the study repeatedly through different pregnancies, we used a robust sandwich covariance matrix estimate to account for interdependent observations. Level of significance was set as $p = 0.05$ (two-sided) and statistical analyses were performed using SAS version 9.2 (SAS Institute Inc., Cary, NC, USA). No adjustment for multiple testing was performed.

2.5. Ethics

The Regional Scientific Ethics Committee for the municipalities of Copenhagen and Frederiksberg approved all DNBC study protocols. Approval from the Danish Data Protection Agency was also obtained. All procedures were in accordance with the Declaration of Helsinki. All women provided written informed consent.

3. Results

3.1. Basic Characteristics

The characteristics of the study participants are presented in Table 1. We confirm a higher percentage with high socio-economic status among those who take iron during pregnancy and give iron supplementation to their child. Whereas maternal celiac disease and anemia seems to influence the likelihood of supplementation during early pregnancy, but not supplementation to the child.

Table 1. Maternal characteristics in relation to maternal use of pure iron supplements and offspring intake of iron droplets during the first 18 months of infancy.

	All	Maternal Use of Iron Supplements			Offspring Use of Iron Droplets	
		No Use (n = 15,071)	Early Use (n = 11,092)	Late Use (n = 48,318)	No (n = 24,281)	Yes (n = 27,595)
Maternal age (years)	30.4 (4.2)	30.7 (4.4)	30.7 (4.2)	30.3 (4.1)	30.5 (4.2)	30.6 (4.1)
Pre-pregnancy BMI (kg/m^2)						
% underweight (<18.5)	4.3%	4.1%	5.4%	4.2%	3.9%	4.2%
% normal weight (18.5–25)	68.2%	66.1%	70.5%	68.8%	66.4%	69.7%
% overweight (25–30)	19.6%	20.6%	17.8%	19.4%	20.9%	18.8%
% obese (>30)	7.9%	9.2%	6.4%	7.6%	8.8%	7.3%
Maternal Smoking, %	24.5%	27.2%	25.5%	23.4%	26.0%	21.5%
Nulliparous, %	48.4%	39.5%	41.3%	51.5%	45.5%	49.8%
Socio-economic status						
High, %	55.3%	52.7%	54.9%	56.3%	52.6%	58.9%
Medium, %	27.2%	27.9%	28.0%	27.1%	29.8%	25.4%
Low, %	12.2%	14.3%	12.7%	11.3%	13.0%	10.3%
Students, %	5.3%	5.1%	4.5%	5.3%	4.6%	5.3%
Breastfeeding, %						
No	11.1%	11.6%	11.1%	10.8%	13.2%	6.3%
1–6 months	28.4%	28.9%	28.8%	28.1%	33.3%	17.5%
6+ months	60.6%	59.5%	60.1%	61.1%	53.5%	76.2%
Gestational age at delivery (days)	280.2 (12.4)	280.8 (11.5)	280.9 (21.9)	280.8 (11.2)	281.1 (10.8)	279.8 (13.0)
Cesarean section, %	15.3%	14.1%	13.8%	15.2%	15.1%	15.1%
Maternal celiac disease (n,%)	172, 0.25%	36, 0.24%	45, 0.41%	122, 0.25%	57, 0.23%	64, 0.24%
Maternal type 1 diabetes (n,%)	332, 0.49%	77, 0.51%	44, 0.40%	232, 0.48%	114, 0.47%	137, 0.53%
Maternal anemia (n,%)	1916, 3.5%	427, 3.6%	606, 7.0%	1374, 3.5%	732, 3.6%	804, 3.5%

Early use = before gestational week 20; late use = after gestational week 20; Body Mass Index, BMI.

The median age at T1D diagnosis was 9.8 years (range 0.9–16.9), and the median follow-up time for the cohort sample was 15.6 years (range 13.0–18.6). Maternal intake of any iron supplement was below 40 mg/daily and above 80 mg/daily in 38.2% and 4.6% of the women, respectively. The proportion of women reporting to take pure iron supplement during early (<20 weeks) and late (≥20 weeks) pregnancy was 17.3% and 75,6%, respectively. The proportion of children receiving iron droplets during the first 18 months of life was 53.2%—the majority of these children (92.3%) were only supplemented with iron droplets for six months or less.

3.2. Maternal Iron Supplementation

Our primary adjusted analysis demonstrated that there was no association between maternal pure iron supplementation during pregnancy and later risk of childhood T1D in the offspring HR 1.05 (95% CI: 0.76–1.45). Further, maternal pure iron supplementation was not associated with risk of T1D in the offspring prior to or after gestational week 20: HR 0.82 (95% CI: 0.57–1.17) and HR 1.13 (95% CI: 0.83–1.53), respectively (Table 2). Further, no association was found when categorizing maternal iron intake from any supplement in mid-pregnancy and later risk of T1D in the offspring (p_{trend} = 0.82) (Table 2).

Table 2. Associations between maternal intake of pure iron supplements during pregnancy and total intake of iron from supplements in relation to offspring risk of type 1 diabetes.

	No. Cases (%)/N	Unadjusted	Adjusted 1 [1]	Adjusted 2 [2]
Any use of pure iron supplements (*n* = **63,931**)				
No	**48 (0.36%)**/13,196	1.00 (reference)	1.00 (reference)	1.00 (reference)
Yes	**190 (0.37%)**/50,735	1.06 (0.77, 1.46)	1.05 (0.76, 1.45)	1.05 (0.77, 1.45)
Early use of pure iron supplements in gestational week 1 to 19 (*n* = **63,931**)				
No	**203 (0.38%)**/52,834	1.00 (reference)	1.00 (reference)	1.00 (reference)
Yes–early	**35 (0.32%)**/11,097	0.81 (0.57, 1.16)	0.82 (0.57, 1.17)	0.81 (0.57, 1.16)
Late use of pure iron supplements in gestational week 20 to 40 (*n* = **63,931**)				
No	**54 (0.35%)**/15,613	1.00 (reference)	1.00 (reference)	1.00 (reference)
Yes–late	**184 (0.38%)**/48,318	1.14 (0.84, 1.54)	1.13 (0.83, 1.53)	1.13 (0.83, 1.54)
Total supplemental iron intake as reported in week 25 of gestation (*n* = **68,240**)				
0 mg/day	**29 (0.42%)**/6880	1.00 (reference)	1.00 (reference)	1.00 (reference)
>0–20 mg/day	**12 (0.28%)**/4308	0.66 (0.34, 1.29)	0.67 (0.34, 1.32)	0.66 (0.34, 1.31)
>20–40 mg/day	**51 (0.34%)**/14,934	0.81 (0.51, 1.27)	0.80 (0.50, 1.26)	0.79 (0.50, 1.25)
>40–60 mg/day	**117 (0.40%)**/29,539	0.95 (0.63, 1.43)	0.94 (0.62, 1.43)	0.94 (0.86, 1.41)
>60–80 mg/day	**37 (0.39%)**/9431	0.94 (0.58, 1.53)	0.94 (0.58, 1.53)	0.93 (0.57 1.52)
>80 mg/day	**11 (0.35%)**/3148	0.81 (0.41, 1.63)	0.81 (41, 1.63)	0.80 (0.40, 1.60)
p-value for trend		0.80	0.82	0.84

Number and percentage of cases are depicted in bold. [1] Adjusted for parental socio-economic status, mode of delivery, pre-pregnancy BMI, age, smoking status, parity, gestational age, maternal age, and breastfeeding. [2] same as model 1 but in addition adjustments are made for maternal celiac disease, maternal type 1 diabetes and maternal anemia.

3.3. Infant Iron Supplementation

We found an inverse association between iron supplementation during the first 18 months of life and risk of childhood T1D HR 0.74 (95% CI: 0.55–1.00) (p_{trend} = 0.03) (Table 3) (Figure 1).

Table 3. Associations between offspring intake of iron droplets during first 18 months of life in relation to later risk of type 1 diabetes.

	No. Cases (%)/N	Unadjusted	Adjusted 1 [1]	Adjusted 2 [2]
Offspring use of iron droplets reported at 18 months (n = **51,859**)				
No	**104 (0.43%)**/24,272	1.00 (reference)	1.00 (reference)	1.00 (reference)
Yes	**87 (0.32%)**/27,587	0.73 (0.55, 0.97)	0.74 (0.55, 1.00)	0.73 (0.55, 0.99)
No	**104 (0.43%)**/24,272	1.00 (reference)	1.00 (reference)	1.00 (reference)
1–6 months	**82 (0.32%)**/25,483	0.75 (0.56, 1.00)	0.76 (0.56, 1.02)	0.75 (0.55, 1.01)
>6 months	**5 (0.24%)**/2104	0.56 (0.28, 1.37)	0.55 (0.23, 1.35)	0.56 (0.23, 1.36)
p-value for trend [3]		0.03	0.03	0.03
p-value for effect [4]		0.08	0.10	0.09

Number and percentage of cases are depicted in bold. [1] Adjusted for parental socio-economic status, mode of delivery, pre-pregnancy BMI, age, smoking status, parity, gestational age, maternal age, and breastfeeding. [2] Same as model 1 but in addition adjustments are made for maternal celiac disease, maternal type I diabetes and maternal anemia. [3] Chi-square-test is used to test for a linear dose-response—the iron variable is included in the regression model as a continuous (three values) variable. [4] Chi-square-test is used to test the null hypothesis that all three groups are equal.

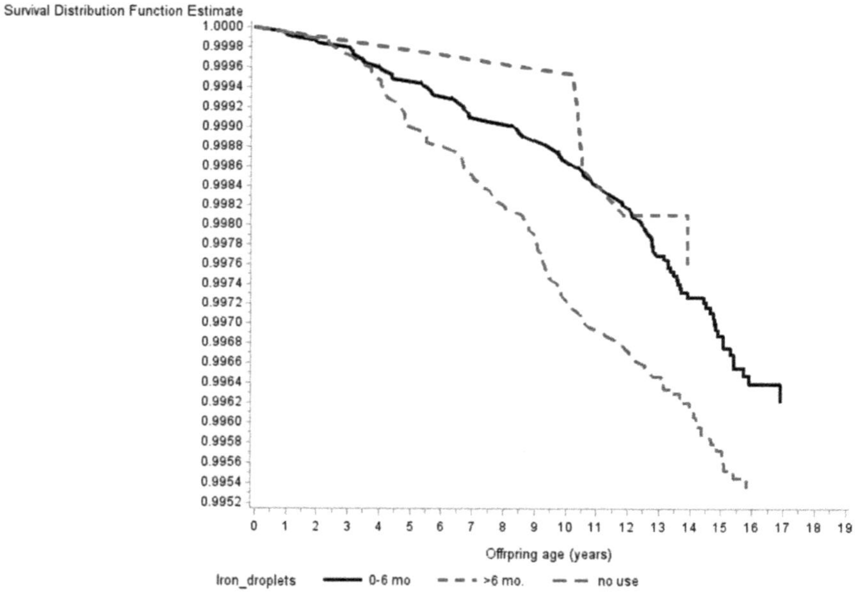

Figure 1. Survival curve illustrating the inverse association between early life supplementation with iron droplets and the risk of childhood type 1 diabetes. p-values for trend and effect are seen in Table 3. Note that the y-axis does not begin at zero.

3.4. Sensitivity Analyses

Our results were essentially unchanged when additional adjustment for maternal anaemia, maternal celiac disease, and maternal T1D were performed (Tables 2 and 3).

When we start follow-up at 30 months of age (one year after information of iron droplets was collected) no differences are seen in our results regarding offspring risk of T1D.

4. Discussion

In this study, which utilizes one of the world's largest cohorts of pregnant women, we show that maternal iron intake through supplementation is not associated with later risk of childhood T1D in the offspring, but iron supplementation in early childhood may be protective against childhood T1D.

4.1. Comparison with Other Studies

4.1.1. Studies Regarding Maternal Iron Intake and Later Risk of Childhood T1D

To our knowledge only the prospective Norwegian Mother and Child Cohort Study has examined the relationship between maternal iron supplementation and status, and later risk of T1D in the offspring [23]. They found a positive association between any maternal iron supplementation and childhood T1D (adjusted HR 1.33 (95% CI: 1.06–1.67)), when only supplements containing pure iron or iron in combination with other micronutrients were examined the results were no longer statistically significant (adjusted HR 1.71 (95% CI: 0.83–3.53)) and (adjusted HR 1.60 (95% CI: 0.95–2.69)), respectively. The Norwegian study found no association between maternal intake of iron from food and later risk of T1D in the offspring.

4.1.2. Iron Supplementation Guidelines in Denmark and Norway during the Recruitment Period

Størdal and colleagues' states that iron supplementation of pregnant women was during the first half of their recruitment period (2000–2005) based on national guidelines concerning ferritin measurements. Revised guidelines did not recommend serum ferritin measurements, but rather recommended haemoglobin measurement at week 30. However, analyses revealed iron supplementation was given as a broad recommendation (40 mg/d from gestational week 18–20) during the whole study period, and to a lesser extend guided by indices of body iron stores [23]. During the DNBC recruitment period all Danish pregnant women were advised to take a daily supplement of 50–70 mg iron from week 20 of gestation until delivery [28]. In the DNBC study, 37% of Danish pregnant women had an iron supplementation intake above the recommended 50–70 mg per day, due to inappropriate product formulations on the Danish market [16]. This indicates that Danish pregnant women, during the overlapping recruitment periods, had a higher supplemental iron exposure compared to the Norwegian women, but still we did not find an association between supplemental iron exposure and risk of T1D in the offspring. Iron supplements were used at some point in 64% of the women in the Norwegian study versus 79% of the Danish women that took pure iron supplements at some point during pregnancy [23].

4.1.3. Studies Regarding Infant Iron Intake and Later Risk of Childhood T1D

The Norwegian Mother and Child Cohort Study found no association between early life (<18 months of life) iron supplementation and risk of childhood T1D (adjusted HR 1.22 (95% CI: 0.78–1.90)). Infant iron supplementation in Norway was not common practice during the study period. At six and 18 months only 4.3% and 1.4% of the Norwegian infants received an iron supplement, respectively, which leaves this study with insufficient power to detect a possible effect [23]. In the DNBC study 49.1% of the infants received iron droplets during a period of one to six months, but only 4.1% received iron droplets longer than six months (Table 3). During the DNBC study recruitment period the Danish Health Authority recommended that children were iron supplemented between six and 12 months.

In addition, only five retrospective studies have looked at iron exposure before the age of 16 years and risk of developing childhood T1D [25,29–32]. The results are inconsistent but there are important methodological differences and limitations to take into account. A total of three studies, one retrospective and two case-control studies, focused on the content and amount of trace metals, including iron, in drinking water at the time of T1D diagnosis. None of these studies found statistically significant associations, but the study by Samuelson and colleagues suggested that a high iron content in the drinking water increases the risk of childhood T1D odds ratio (OR) 1.56 (95% CI: 0.99–2.44). These studies reported iron concentrations in drinking water of 0.01–0.08 mg/L, which contributes to a negligible iron intake compared to iron gained from supplementation and diet [33]. Moreover, iron exposure around time of childhood T1D diagnosis represents another "window of vulnerability" compared to the present study's exposure interval i.e., in utero to 18 months of age. Further, one case-control study found that the odds ratio for one SD increase in iron intake (from infant formulas and breast milk) was 2.01 (95% CI: 1.18–3.41) during the first four months of life [32]. Iron exposure was quantified from a self-administered questionnaire filled out by parents of children with T1D under the age of 10 years. Children were diagnosed with T1D between one and six years of age. Using self-reported questionnaires on dietary recall over a long duration may introduce differential misclassification of exposure i.e., mothers of control subjects may under-report exposure and mothers of case subjects may over-report, which in both situations leads to overestimation of the association [34]. Lastly, a Danish case-control study, originated from our research group, quantified neonatal whole blood (WB-Iron) iron content. For each doubling in WB-Iron content the odds of developing childhood T1D was 2.55 (95% CI: 1.04–6.24). This study only had one biological measurement and no information on iron homeostasis genetics, but a direct quantification of total iron content using a valid method seems superior to questionnaire data on iron intake, due to variation in bioavailability of iron from both the food source and supplements [35,36], competing mechanisms at the di-metal luminal transporters, and genetic factors regulating the iron transport from the intestinal lumen to the circulation [24,37] However, the measured WB-Iron content at birth is likely to reflect the intrauterine exposure and not early life supplementation.

Our finding that supplemental iron intake during pregnancy is not associated with childhood T1D in the offspring, may primarily be due to: i. a tight placental iron regulation that is affected by dietary/supplemental iron intake and the iron metabolism genetic make-up in both the mother and the fetus [38]; ii. difference in intestinal absorption depending on supplemental-regime e.g., magnitude of hepcidin counterbalance when iron supplements are taken consecutively versus alternately [36]; and iii. no effect of iron on risk of childhood T1D. Regarding the placental iron regulation, neonates born by women that had been supplemented with 66 mg elemental iron daily during pregnancy had s-ferritin concentrations that were higher compared to neonates born of women that did not receive an iron supplement during pregnancy (155 versus 118 microgram/L). Neonates born by women with s-ferritin below 13.6 microgram/L also have lower levels of s-ferritin [19]. S-ferritin < 20 microgram/L indicates exhausted iron stores [39]. The Norwegian Mother and Child Cohort Study only found a suggestive non-significant positive association between s-ferritin in cord blood and later risk of childhood T1D (adjusted odds ratio 1.05 (95% CI: 0.99–1.13) per 50 mg/L increase) [23]. The 5th–95th percentile reference interval for cord s-ferritin was 40–468 microgram/L in Norway [40]. Hay et al. states that cord s-ferritin is a strong predictor of iron status the first two years of life [40].

4.2. Strengths and Weaknesses

Our study benefits from several strengths: (i) a relative large-scale study with a prospective design; (ii) we were able to obtain information on iron intake from both pregnancy and early childhood; and (iii) multiple possible confounders were included in our analyses.

Some limitation should also be considered: (i) we have no genetic data regarding iron homeostasis; (ii) using iron intake as *a proxy* for body iron status is inferior to direct quantification [15], but the fact that iron supplementation is associated with iron-status biomarkers in both pediatric and adult populations [41,42], and iron requirements fall within a relatively narrow band, with low and high intakes being harmful, makes data on iron intake, especially from large prospective cohort, a valuable source for gaining new insight within this field of study; (iii) participants in the DNBC study may not be representative of the general population of pregnant women in Denmark (e.g., they may have a healthier lifestyle and be better educated), but this does not necessarily confound exposure-outcome associations [43]; (iv) users of dietary supplements in general have a healthier diet than non-users, and individuals who take dietary supplements are the most unlikely to need them [16], which, in the DNBC study, results in ~37% of the women being exposed to iron supplements intakes that exceeds the recommended daily dose of 50–70 mg during pregnancy; (v) our results may be generalizable to other European and European origin populations, but may not be generalizable to populations with higher prevalence of iron deficiency and iron deficiency anemia; (vi) very few women did not take iron supplementation in Denmark decreasing the power leaving the 33% increased risk found in the Norwegian study within our 95% CI; (vii) we did not have any data on development of persistent islet autoantibodies in the children, prior to T1D diagnosis, and were therefore not able to examine the effect of maternal and offspring iron supplementation on the risk of initiation (seroconversion) and/or acceleration (epitope spreading) of islet autoimmunity. Our start of follow-up was 18 months, in the offspring analyses, excluding children diagnosed before this age (1.56%) and no change in results were seen when follow-up was postponed one year (30 months of age); (viii) potential micronutrient-micronutrient interactions regarding association with childhood T1D could not be well-examined in this study. Though, there is no strong evidence in the literature that iron supplementation would e.g., affect zinc status negatively or vice versa or that low zinc status could increase risk of T1D [44–46]; and (ix) our results may be influenced by residual confounding.

4.3. Implications and Future Perspective

In contrast with previous reports we found no indication that maternal use of iron supplements during pregnancy is harmful with respect to later risk of developing childhood T1D in the offspring. In addition, infant use of iron droplets during the first 18 months of life, was if anything, protective with respect to later T1D risk, but these findings need to be confirmed before strong conclusions can be reached. Future prospective studies need to be large-scale and must preferable include sensitive iron status markers and a broad-range of iron homeostasis single nucleotide polymorphisms for a thorough understanding of complex iron-genotype interactions, which could shed light on possible at-risk/protected groups.

5. Conclusions

Our large-scale prospective study demonstrated no harmful effects of iron supplementation during pregnancy and in early life in regard to later risk of childhood T1D in the offspring.

Author Contributions: Conceptualization, S.U.T., T.I.H., A.A.B., S.F.O. and J.S.; methodology, S.U.T., T.I.H., S.F.O. and J.S.; software, T.I.H; validation, T.I.H; investigation, S.U.T., T.I.H., A.A.B., S.F.O. and J.S.; resources, T.I.H., S.F.O. and J.S.; data curation, T.I.H; writing—original draft, S.U.T., T.I.H., A.A.B., S.F.O. and J.S.; writing—review and editing, S.U.T., T.I.H., A.A.B., S.F.O. and J.S.; visualization, T.I.H.; supervision, S.U.T., T.I.H. and J.S.; projection administration, S.U.T., T.I.H. and J.S.; funding acquisition, S.U.T., T.I.H., S.F.O. and J.S.

Funding: The Danish National Birth Cohort Study was supported by the Danish Heart Association, Danish Medical Research Council, Sygekassernes Helsefond and the Danish National Research Foundation, and its dietary component was supported by Innovation Fund Denmark (grant No 09-067124, Centre for Fetal Programming), the March of Dimes Birth Defects Foundation (6-FY-96-0240, 6-FY97-0553, 6-FY97-0521, 6-FY00-407).

Conflicts of Interest: The authors declare no conflict of interest. The funders had no role in the design of the study; in the collection, analyses, or interpretation of data; in the writing of the manuscript, or in the decision to publish the results.

References

1. Atkinson, M.A.; Eisenbarth, G.S.; Michels, A.W. Type 1 diabetes. *Lancet* **2014**, *383*, 69–82. [CrossRef]
2. Eizirik, D.L.; Colli, M.L.; Ortis, F. The role of inflammation in insulitis and β-cell loss in type 1 diabetes. *Nat. Rev. Endocrinol.* **2009**, *5*, 219–226. [CrossRef]
3. Pociot, F.; Lernmark, Å. Genetic risk factors for type 1 diabetes. *Lancet* **2016**, *387*, 2331–2339. [CrossRef]
4. Rewers, M.; Ludvigsson, J. Environmental risk factors for type 1 diabetes. *Lancet* **2016**, *387*, 2340–2348. [CrossRef]

5. Svensson, J.; Lyngaae-Jørgensen, A.; Carstensen, B.; Simonsen, L.B.; Mortensen, H.B. Long-term trends in the incidence of type 1 diabetes in Denmark: The seasonal variation changes over time. *Pediatr. Diabetes* **2009**, *10*, 248–254. [CrossRef]
6. Patterson, C.C.; Dahlquist, G.G.; Gyürüs, E.; Green, A.; Soltész, G. Incidence trends for childhood type 1 diabetes in Europe during 1989–2003 and predicted new cases 2005–20: A multicentre prospective registration study. *Lancet* **2009**, *373*, 2027–2033. [CrossRef]
7. Das, I.; Saha, K.; Mukhopadhyay, D.; Roy, S.; Raychaudhuri, G.; Chatterjee, M.; Mitra, P.K. Impact of iron deficiency anemia on cell-mediated and humoral immunity in children: A case control study. *J. Nat. Sci. Biol. Med.* **2014**, *5*, 158–163.
8. Attia, M.A.; Essa, S.A.; Nosair, N.A.; Amin, A.M.; El-Agamy, O.A. Effect of iron deficiency anemia and its treatment on cell mediated immunity. *Indian J. Hematol. Blood Transfus.* **2009**, *25*, 70–77. [CrossRef]
9. Sampaio, A.F.S.; Silva, M.; Dornas, W.C.; Costa, D.C.; Silva, M.E.; dos Santos, R.C.; de Lima, W.G.; Pedrosa, M.L. Iron toxicity mediated by oxidative stress enhances tissue damage in an animal model of diabetes. *Biometals* **2014**, *27*, 349–361. [CrossRef]
10. Backe, M.B.; Moen, I.W.; Ellervik, C.; Hansen, J.B.; Mandrup-Poulsen, T. Iron Regulation of Pancreatic Beta-Cell Functions and Oxidative Stress. *Annu. Rev. Nutr.* **2016**, *36*, 241–273. [CrossRef]
11. Tiedge, M.; Lortz, S.; Drinkgern, J.; Lenzen, S. Relation between antioxidant enzyme gene expression and antioxidative defense status of insulin-producing cells. *Diabetes* **1997**, *46*, 1733–1742. [CrossRef] [PubMed]
12. Lenzen, S. Oxidative stress: The vulnerable beta-cell. *Biochem. Soc. Trans.* **2008**, *36*, 343–347. [CrossRef]
13. Lee, A.I.; Okam, M.M. Anemia in pregnancy. *Hematol. Oncol. Clin. North Am.* **2011**, *25*, 241–259. [CrossRef]
14. Milman, N. Iron in pregnancy: How do we secure an appropriate iron status in the mother and child? *Ann. Nutr. Metab.* **2011**, *59*, 50–54. [CrossRef] [PubMed]
15. Van der Merwe, L.F.; Eussen, S.R. Iron status of young children in Europe. *Am. J. Clin. Nutr.* **2017**, *106*, 1663S–1671S. [CrossRef]
16. Knudsen, V.K.; Hansen, H.S.; Ovesen, L.; Mikkelsen, T.B.; Olsen, S.F. Iron supplement use among Danish pregnant women. *Public Health Nutr.* **2007**, *10*, 1104–1110. [CrossRef] [PubMed]
17. *Nordic Council of Ministers Nordic Nutrition Recommendations 2012: Integrating Nutrition and Physical Activity*, 5th ed.; Nordic Council of Ministers: Copenhagen, Denmark, 2014.
18. European Food Safety Authority (Ed.) *Tolerable Upper Intake Levels for Vitamins and Minerals*; European Food Safety Authority: Parma, Italy, 2006; ISBN 978-92-9199-014-6.
19. Shao, J.; Lou, J.; Rao, R.; Georgieff, M.K.; Kaciroti, N.; Felt, B.T.; Zhao, Z.-Y.; Lozoff, B. Maternal Serum Ferritin Concentration Is Positively Associated with Newborn Iron Stores in Women with Low Ferritin Status in Late Pregnancy. *J. Nutr.* **2012**, *142*, 2004–2009. [CrossRef]
20. Türkay, S.; Tanzer, F.; Gültekin, A.; Bakici, M.Z. The influence of maternal iron deficiency anaemia on the haemoglobin concentration of the infant. *J. Trop. Pediatr.* **1995**, *41*, 369–371. [CrossRef]
21. Balesaria, S.; Hanif, R.; Salama, M.F.; Raja, K.; Bayele, H.K.; McArdle, H.; Srai, S.K.S. Fetal iron levels are regulated by maternal and fetal Hfe genotype and dietary iron. *Haematologica* **2012**, *97*, 661–669. [CrossRef]
22. Saffery, R.; Novakovic, B. Epigenetics as the mediator of fetal programming of adult onset disease: What is the evidence? *Acta Obs. Gynecol. Scand.* **2014**, *93*, 1090–1098. [CrossRef]
23. Størdal, K.; McArdle, H.J.; Hayes, H.; Tapia, G.; Viken, M.K.; Lund-Blix, N.A.; Haugen, M.; Joner, G.; Skrivarhaug, T.; Mårild, K.; et al. Prenatal iron exposure and childhood type 1 diabetes. *Sci. Rep.* **2018**, *8*, 9067. [CrossRef]
24. Søgaard, K.L.; Ellervik, C.; Svensson, J.; Thorsen, S.U. The Role of Iron in Type 1 Diabetes Etiology: A Systematic Review of New Evidence on a Long-Standing Mystery. *Rev. Diabet. Stud.* **2017**, *14*, 269–278. [CrossRef]
25. Kyvsgaard, J.N.; Overgaard, A.J.; Thorsen, S.U.; Hansen, T.H.; Pipper, C.B.; Mortensen, H.B.; Pociot, F.; Svensson, J. High Neonatal Blood Iron Content Is Associated with the Risk of Childhood Type 1 Diabetes Mellitus. *Nutrients* **2017**, *9*, 1221. [CrossRef] [PubMed]

26. Olsen, J.; Melbye, M.; Olsen, S.F.; Sorensen, T.I.A.; Aaby, P.; Andersen, A.-M.N.; Taxbol, D.; Hansen, K.D.; Juhl, M.; Schow, T.B.; et al. The Danish National Birth Cohort—Its background, structure and aim. *Scand. J. Public Health* **2001**, *29*, 300–307. [CrossRef]
27. Svensson, J.; Cerqueira, C.; Kjærsgaard, P.; Lyngsøe, L.; Hertel, N.T.; Madsen, M.; Mortensen, H.B.; Johannesen, J. Danish Registry of Childhood and Adolescent Diabetes. *Clin. Epidemiol.* **2016**, *8*, 679. [CrossRef]
28. Sundhedsstyrelsen (National Board of Health). *Svangreomsorg: Retningslinier og Redegørelse*; Guidelines for Prenatal Care; Sundhedsstyrelsen: Copenhagen, Denmark, 1998.
29. Samuelsson, U.; Oikarinen, S.; Hyöty, H.; Ludvigsson, J. Low zinc in drinking water is associated with the risk of type 1 diabetes in children. *Pediatr. Diabetes* **2011**, *12*, 156–164. [CrossRef] [PubMed]
30. Winkler, C.; Mollenhauer, U.; Hummel, S.; Bonifacio, E.; Ziegler, A.-G. Exposure to environmental factors in drinking water: Risk of islet autoimmunity and type 1 diabetes—The BABYDIAB study. *Horm. Metab. Res.* **2008**, *40*, 566–571. [CrossRef]
31. Zhao, H.X.; Mold, M.D.; Stenhouse, E.A.; Bird, S.C.; Wright, D.E.; Demaine, A.G.; Millward, B.A. Drinking water composition and childhood-onset Type 1 diabetes mellitus in Devon and Cornwall, England. *Diabet. Med.* **2001**, *18*, 709–717. [CrossRef] [PubMed]
32. Ashraf, A.P.; Eason, N.B.; Kabagambe, E.K.; Haritha, J.; Meleth, S.; McCormick, K.L. Dietary iron intake in the first 4 months of infancy and the development of type 1 diabetes: A pilot study. *Diabetol. Metab. Syndr.* **2010**, *2*, 58. [CrossRef]
33. *Institute of Medicine (US) Panel on Micronutrients Dietary Reference Intakes for Vitamin A, Vitamin K, Arsenic, Boron, Chromium, Copper, Iodine, Iron, Manganese, Molybdenum, Nickel, Silicon, Vanadium, and Zinc*; National Academies Press: Washington, DC, USA, 2001; ISBN 978-0-309-07279-3.
34. Rothman, K.; Greenland, S.L.T. *Modern Epidemiology*, 3rd ed.; Lippincott Publishers: Philadelphia, PA, USA, 2008.
35. Hurrell, R.; Egli, I. Iron bioavailability and dietary reference values. *Am. J. Clin. Nutr.* **2010**, *91*, 1461S–1467S. [CrossRef] [PubMed]
36. Stoffel, N.U.; Cercamondi, C.I.; Brittenham, G.; Zeder, C.; Geurts-Moespot, A.J.; Swinkels, D.W.; Moretti, D.; Zimmermann, M.B. Iron absorption from oral iron supplements given on consecutive versus alternate days and as single morning doses versus twice-daily split dosing in iron-depleted women: Two open-label, randomised controlled trials. *Lancet Haematol.* **2017**, *4*, e524–e533. [CrossRef]
37. Anderson, G.J.; Frazer, D.M. Current understanding of iron homeostasis. *Am. J. Clin. Nutr.* **2017**, *106*, 1559S–1566S. [CrossRef]
38. Gambling, L.; Lang, C.; McArdle, H.J. Fetal regulation of iron transport during pregnancy. *Am. J. Clin. Nutr.* **2011**, *94*, 1903S–1907S. [CrossRef]
39. Milman, N.; Agger, A.O.; Nielsen, O.J. Iron supplementation during pregnancy. Effect on iron status markers, serum erythropoietin and human placental lactogen. A placebo controlled study in 207 Danish women. *Dan. Med. Bull.* **1991**, *38*, 471–476.
40. Hay, G.; Refsum, H.; Whitelaw, A.; Melbye, E.L.; Haug, E.; Borch-Iohnsen, B. Predictors of serum ferritin and serum soluble transferrin receptor in newborns and their associations with iron status during the first 2 y of life. *Am. J. Clin. Nutr.* **2007**, *86*, 64–73. [CrossRef]
41. Casgrain, A.; Collings, R.; Harvey, L.J.; Hooper, L.; Fairweather-Tait, S.J. Effect of iron intake on iron status: A systematic review and meta-analysis of randomized controlled trials. *Am. J. Clin. Nutr.* **2012**, *96*, 768–780. [CrossRef]
42. Athe, R.; Rao, M.V.V.; Nair, K.M. Impact of iron-fortified foods on Hb concentration in children (<10 years): A systematic review and meta-analysis of randomized controlled trials. *Public Health Nutr.* **2014**, *17*, 579–586.
43. Nohr, E.A.; Frydenberg, M.; Henriksen, T.B.; Olsen, J. Does low participation in cohort studies induce bias? *Epidemiology* **2006**, *17*, 413–418. [CrossRef]
44. Fischer Walker, C.; Kordas, K.; Stoltzfus, R.J.; Black, R.E. Interactive effects of iron and zinc on biochemical and functional outcomes in supplementation trials. *Am. J. Clin. Nutr.* **2005**, *82*, 5–12. [CrossRef]

45. Uusitalo, L.; Kenward, M.G.; Virtanen, S.M.; Uusitalo, U.; Nevalainen, J.; Niinistö, S.; Kronberg-Kippilä, C.; Ovaskainen, M.-L.; Marjamäki, L.; Simell, O.; et al. Intake of antioxidant vitamins and trace elements during pregnancy and risk of advanced beta cell autoimmunity in the child. *Am. J. Clin. Nutr.* **2008**, *88*, 458–464. [CrossRef]
46. Kyvsgaard, J.N.; Overgaard, A.J.; Jacobsen, L.D.; Thorsen, S.U.; Pipper, C.B.; Hansen, T.H.; Husted, S.; Mortensen, H.B.; Pociot, F.; Svensson, J. Low perinatal zinc status is not associated with the risk of type 1 diabetes in children. *Pediatr. Diabetes* **2017**, *18*, 637–642. [CrossRef] [PubMed]

nutrients

MDPI

Article

The Safety and Tolerability of a Potential Alginate-Based Iron Chelator; Results of A Healthy Participant Study

Richard D. Horniblow [1], Pritesh Mistry [1,2], Mohammed N. Quraishi [1,3], Andrew D. Beggs [1,2], Tom Van de Wiele [4], Tariq H. Iqbal [1,2,3] and Chris Tselepis [1,*]

1 Institute of Cancer and Genomic Sciences, University of Birmingham, Edgbaston, Birmingham B15 2TT, UK; r.horniblow@bham.ac.uk (R.D.H.); P.Mistry.1@bham.ac.uk (P.M.); M.N.Quraishi@bham.ac.uk (M.N.Q.); A.Beggs@bham.ac.uk (A.D.B.); T.H.IQBAL@bham.ac.uk (T.H.I.)

2 Queen Elizabeth Hospital Birmingham, Edgbaston, Birmingham B15 2TH, UK

3 The University of Birmingham Microbiome Treatment Centre, University of Birmingham, Birmingham B15 2TT, UK

4 CMET, Center for Microbial Ecology and Technology, Faculty of Bioscience Engineering, Ghent University, Coupure Links 653, B-9000 Gent, Belgium; Tom.VandeWiele@UGent.be

* Correspondence: c.tselepis@bham.ac.uk; Tel.: +44-0121-414-8665

Received: 31 January 2019; Accepted: 13 March 2019; Published: 21 March 2019

Abstract: Evidence supporting the ferro-toxic nature of iron in the progression of inflammatory bowel disease (IBD) is becoming well established. A microbial dysbiosis is observed in IBD patients, and intra-luminal colonic-iron is able to support a more pathogenic community of bacteria; whether this is attributed to the development of IBD and how iron could be mediating these microbial changes is still unknown. Dietary fibres are commonly used in pre-biotic supplements to beneficially affect the host by improving the viability of bacterial communities within the colon. Alginates are a class of biopolymers considered as prebiotics due to their fibre-like composition and are able to bind metal cations, in particular, iron. Considering that iron excess is able to negatively alter the microbiome, the use of alginate as a food supplement could be useful in colonic-iron chelation. As such, this first-in-man study aimed to assess whether the use of alginate as a dietary iron chelator was both safe and well tolerated. In addition, the impact of alginate on the microbiome and iron levels was assessed by using an intestinal model SHIME (Simulation of the Human Intestinal Microbial Ecosystem). Alginate was supplemented into the diets (3 g/day) of healthy volunteers (n = 17) for 28 days. Results from this study suggest that daily ingestion of 3 g alginate was well tolerated with very minor side effects. There were no detrimental changes in a variety of haematological parameters or the intestinal microbiome. The bacterial communities within the SHIME model were also not influenced by iron and or alginate; it is possible that alginate may be susceptible to bacterial or enzymatic degradation within the gastro-intestinal tract.

Keywords: alginate; iron; microbiome; ROS; prebiotic

1. Introduction

It has recently been highlighted that elevated concentrations of iron within the colon are associated with the progression and development of intestinal diseases, and specifically inflammatory bowel disease (IBD) [1–3]. This is supported by the finding that depleting iron within the colon is associated with an ameliorated phenotype [4]. Additionally, enzymes responsible for the breakdown of iron-complexes (heme-oxygenases) were found to be upregulated and protective in instances of inflammation within the colon, possibly in an attempt to diminish iron's pro-inflammatory nature [5]. High levels of colonic luminal iron have been shown to (i) be pro-inflammatory [6,7], (ii) have catalytic reactive oxygen species (ROS) generating ability [6,8], (iii) negatively alter the microbiome [1,9–13] and (iv) have cellular proliferation effects [14,15]. When dietary iron consumption is in excess of nutritional demands (average consumption of 10 to 12 mg with only 2 mg required) and only 0.7 to 22.9% of ingested iron is absorbed [16], this suggests a high residual concentration of iron in the colon. This is particularly the case for patients with IBD who often present with iron-deficiency anaemia and are on oral iron supplements [17,18].

An intervention that would render this excess colonic iron inert would theoretically be highly beneficial; a class of bio-actives that may achieve this would be non-absorbable iron chelators. To date, there have been a number of reports that detail the iron-chelating effects of potentially non-digestible fibres, of which sodium alginate is of particular interest [19–21]. Alginates are natural polysaccharides extracted from the cell walls of brown seaweeds. Chemically, these biopolymers are formed from unbranched (1–4) linked β-D-mannuronic acid (M) and its C5 epimer, α-L-guluronic acid (G) [22]. Alginates are used within the food industry as gelling agents and in anti-reflux preparations due to their ability to gel, forming "rafts" within the stomach to suppress acid-reflux [23]. Most recently, an alginate of relatively short chain length and high M content (MLD) has been identified as an ideal iron chelator, demonstrating limited affinity towards calcium [24]. MLD is a 145 kDa biopolymer with a G:M ratio of 38:62 which is able to bind ferrous iron (0.6 mg Fe/mg MLD) and inhibit murine intestinal iron absorption [24]. Previous clinical studies have demonstrated the ability of alginate supplementation to decrease iron absorption and hence reduce serum-iron concentrations [25]. The use of MLD may be of particular use in the treatment of gastro-intestinal diseases associated with high levels of unabsorbed, luminal iron within the colon. No human studies have been conducted to assess the tolerability of MLD consumption in healthy individuals over a period of 28 days.

Thus, the overall aim of this study was to assess the safety, tolerability and feasibility of 3 g daily alginate consumption in healthy volunteers over a period of 28 days. This was predominantly examined using intestinal health questionnaires, in addition to haematological and faecal microbiome assessments. In a similar approach, the impact of MLD on iron levels and the microbiome was assessed in a highly controlled in vitro model of the human colon. Such results will guide future clinical assessments utilising MLD in iron-chelation therapies.

2. Materials and Methods

2.1. Healthy Participant Study Design

Seventeen healthy individuals, over the age of 18 years of either sex, with no previous history of gastro-intestinal disease were recruited between October 2015 and November 2015. Informed consent was obtained for all participants recruited. Participants completed a quality of life questionnaire (Short Inflammatory Bowel Disease questionnaire (S-IBDQ) (McMaster University, Ontario, Canada) (Figure S1)) and provided a stool sample before consuming MLD. Answers were scored 1–7, with 1 indicating large changes in lifestyle/general wellbeing and 7 indicating no changes or problems. Participants were provided with a diary to record compliance, monitor changes in bowel habit/frequency and document the presence of any adverse symptoms for the duration of the study. Appointments were made for the subsequent two visits required at the midpoint (day 14) and end (day 28) of the study. At each of these visits, further blood and stool samples were collected and S-IBDQ completed. The study design is summarised in Supplementary Figure S2. Venous blood was taken for analysis of full blood count and biochemistry (Figure S3).

2.2. Ethics Statement

The study was approved by the National Research and Ethics Service Committee West Midlands on 20 August 2015 (REC reference: 15/WM/0221). The study was registered at www.isrctn.com as ISRCTN16202716.

2.3. Faecal Microbiota Assessment (Healthy Participant Study and SHIME)

For the assessment of changes in the colonic microbiome, DNA was extracted from stool using the QIAamp DNA stool mini kit (QIAGEN, Hilden, Germany) following the manufacturer's protocol. Three replicate polymerase chain reactions were performed for each faecal sample. In brief, the V4 region of the 16S rRNA gene was amplified with region-specific primers that include the Illumina flowcell adapter sequences. After cluster formation on a HiSeq/MiSeq instrument, the amplicons were sequenced with custom primers.

Quality filtering of reads was applied. Sequences were clustered into operational taxonomic units (OTUs) with 97% similarity and picked using open reference OTU reference picking protocol with a 97% similarity threshold. Taxonomic assignments of OTUs that reached the 97% similarity level were made using QIIME (Quantitative Insights Into Microbial Ecology) by comparison with the Greengene (http://rdp.cme.msu.edu/) databases (gg_13_5_otus). Alpha-diversities of the gut microbial communities were compared by calculating the number of observed OTUs, Shannon Diversity and Faith's Index, and Chao index. Beta diversity was calculated by both unweighted and weighted Unifrac distances calculated. Differences in OTU abundances were tested with ANCOM (Analysis of Composition of Microbiomes). To predict functions encoded by the genomes of bacteria, we performed a phylogenetic investigation of communities by reconstruction of unobserved states (PICRUSt) analysis based on the 16S rRNA analyses. Estimated abundances of Kyoto Encyclopedia of Genes and Genomes (KEGG) Orthology groups were compared between different time points. We obtained 21520246 high-quality filtered reads, corresponding to an average of 377,548 reads per subject. Contaminant reads were negatively filtered, and triplicates were collapsed by the mean value for each observation mean. Reads were clustered into 3525 operational taxonomic units (OTUs) at 97% sequence identity following which their representative sequences were used in taxonomic analysis. All sequences were classified from phylum to species.

2.4. Healthy Volunteer Faecal Iron and Calcium Assessment

Two pools of iron and calcium were assessed; (i) total iron/calcium and (ii) "free" iron/calcium. The total iron/calcium in each sample represents the total metal content, regardless of coordination to ligands/proteins/chemicals whereas the "free" pool represents iron and calcium that is non-associated. To determine the total iron/calcium pool, faecal samples (3 g, wet weight) were ashed in silica crucibles at 480 °C for 48 h within an open-air furnace. The ashed product was re-suspended in 0.1 M HCl (5 mL) and stored. To probe the "free" pool of metal, faecal samples (3 g, wet weight) were homogenised in DI H_2O (3 mL) until a slurry was obtained. An additional volume of DI H_2O (7 mL) was added to the slurry and centrifuged at 7500× *g* for 30 min. The supernatant was collected, and the protein content was removed by acidification. Iron and calcium concentration within the samples was determined using flame atomic absorption spectroscopy (fAAS). Water content of faecal samples was determined by drying overnight at 70 °C and further desiccation.

2.5. Faecal Reactive Oxygen Species Assessment

Faecal samples (2 g, wet weight or 3 mL SHIME colon material) were homogenised and subsequently incubated for 18 h in TRIS buffered saline (4 mL, pH 7.0) containing DMSO (5% (*v*/*v*)), glucose (0.1% (*w*/*v*)) and EDTA (50 mM) under agitation at 37 °C. Faecal slurries were then centrifuged and the supernatant acidified using HCl to precipitate the protein, which was further centrifuged and the resulting supernatant pH re-adjusted to pH = 7.0. Samples were stored at this point at −20 °C until batch analysis could be performed with all samples. Standards were made to run alongside the samples. These were methansulfinic acid (0–10 mM) dissolved in the TRIS buffered used to homogenise the faecal samples; samples and standards were processed identically. A 1.3 mL aliquot of the sample was mixed with H_2SO_4 (10 M, 250 μL) and centrifuged at 3500 RPM for 5 min. The supernatant (1.4 mL) was mixed thoroughly with 1 M sulphuric acid saturated 1-butanol (4 mL) and centrifuged at 3500 RPM for 5 min to allow phase separation. The upper phase was aspirated (3.4 mL) and mixed with sodium acetate (0.5 M, pH 5.0, 2 mL) and again centrifuged at 3500 RPM for 5 min to allow phase separation. The lower aqueous phase was removed (1.8 mL) and pH adjusted to pH 2.5, before the addition of FastBlue BB salt (14 mM, 800 μL) and incubated for 15 min in the dark. After this period, a toluene:butanol (3:1) mixture (1.5 mL) was added and thoroughly mixed for 2 min, before phase separation by centrifugation at 3500 RPM for 5 min. The upper phase (1 mL) was removed and washed with 1-butanol saturated water (2.0 mL). Finally, the samples were centrifuged at 3500 RPM for 5 min, the upper layer (1 mL) was aspirated, mixed with a pyridine:acetic acid (95:5) mixture (100 μL) and absorbance read at λ = 415 nm.

2.6. Intestinal Model (SHIME)

A M-SHIME model was set up as previously described, with proximal colonic vessel attachments only [26]. In brief, the model consisted of a stomach/small intestinal (SI) vessel (not inoculated with faecal bacteria) which fed three independent colonic vessels (inoculated with faecal bacteria) which were regulated at pH 5.6–6.9 (proximal colonic environment). Colonic vessels were adapted with mucin-beads for surface-attached microbes. The beads (K1-carrier, AnoxKaldnes AB, Lund, Sweden) were prepared by boiling mucin and agar in water until they formed a gel and suspended within the colonic vessels.

Each treatment within this experiment (Fe only, MLD+Fe and control) had individual stomach/SI and colonic (n = 3) vessels. Colonic vessels were inoculated, simultaneously with faecal bacteria obtained from one healthy donor. Before supplementing with Fe or MLD, the colonic vessels were equilibrated and established over a 7-day period; between day 7 (t = 0) and 22 (t = 15), the colonic vessels were supplemented with either iron (15 mg) or MLD (1.0 g three times daily). Colonic vessels were flushed daily with N_2 to maintain anaerobicity, and bacterial integrity and health were monitored by measuring short chain fatty acids (SCFAs) and ammonium concentrations regularly. The protocol for these assessments has been previously described [27]. All experiments were run in parallel simultaneously. The whole system was maintained at 37 °C.

The supplementation of aqueous Fe and MLD into SHIME was performed through the addition of stock solutions into the stomach vessel; three injections of $FeSO_4$ in 0.1 M HCl (0.5% w/v, 1 mL) and three injections of MLD (4% w/v, 12.5 mL) per day. The control vessel received DI H_2O only and volumes were adjusted accordingly to ensure each vessel received equal volumes of DI H_2O. These concentrations were pre-optimised to obtain a total Fe concentration within the colonic vessels of c. 400 μM and a free Fe concentration of 100 μM. These concentrations were chosen based on previous in vitro studies examining the iron chelation potential of MLD, and only $FeSO_4$ was used as the iron source in these experiments.

2.7. SHIME Iron Assessments

The measurement of total Fe in the colonic vessels was undertaken by aspirating samples (200 μL) of colonic vessels and mixing with 200 μL of TCA (20% w/v) solution. These were boiled for 5 min at 100 °C and subsequently spun at 1500 RPM for 5 min to pellet debris. The supernatant (200 μL) was aspirated and mixed with a ferrozine solution (600 μL). Measurement of free Fe levels was performed by aspirating 1 mL of colonic medium and spun at 1000 RPM for 1 min to pellet the bacteria. The supernatant (500 μL) was aspirated and directly added to the top of a 30,000 MWCO spin column. This was spun at 15,000 RPM for 10 to 15 min, until equal amounts of filtrate were collected and used directly with the ferrozine reagent (600 μL), as previously described.

2.8. Thin Layer Chromatography (TLC)

TLC was performed to assess the integrity of the alginate following culturing within the colonic vessels of SHIME. A positive control for alginate degradation was prepared by heat degradation as reported previously [24], and degradation derivatives were assessed using a orcinol-based staining reagent, as detailed previously [28].

2.9. Statistical Methods

All analysis was performed using Microsoft Excel. Single factor ANOVA was used to detect changes in the blood results. Paired two-tailed Student's t test was used to detect changes in the levels of bacteria when comparing each of the three time-points in the study at a significance level of 0.05. Values are stated as mean ± standard deviation or median (range).

3. Results

3.1. Study Participant Baseline Demographics

All recruited individuals (13 female and 4 male participants, with a median age of all participants of 42 ± 9.5 years) completed the study. MLD was well tolerated with no serious side effects reported. Minor adverse effects were documented (Table 1), with the predominant complaint of excess flatus followed by bloating.

Table 1. Table of reported adverse effects over the 28-day intervention.

Adverse Effect	Number of Participants Reported
Flatulence	16
Bloating	7
Heartburn	2
Abdominal Pain	8
Loose Motion	4
Urgency to Defecate	1
Nausea and Vomiting	3
Headache	1

There were small but significant changes in S-IBDQ participant survey scores when comparing baseline to day 14 and study endpoint. Mean scores at baseline were 65.5 ± 2.9, which decreased to 62.2 ± 4.0 (p = 0.013) at day 14 and 62.2 ± 4.6 (p = 0.018) at day 28; the total maximum score was 70 (Figure 1A).

Upon consideration of the individual participant responses, these significant changes were associated with questions 4 and 6; 4. How often during the past two weeks have you been troubled by pain in the abdomen? 6. Overall, in the past 2 weeks, how much of a problem have you had with passing large amounts of gas?

The mean response value for question 4 was 6.8, 5.4 and 5.6 at baseline, day 14 and day 28, respectively ($p < 0.005$). These scores equate to a drop from "None of the time" to "A little of the time" or "Hardly any of the time". For question 6 the mean response was 6.6, 4.3 and 4.4 at baseline, day 14 and day 28, respectively ($p < 0.001$). equating to a drop from "No trouble" to "Some trouble". The questionnaire and associated scores are detailed in Supplementary Figure S1.

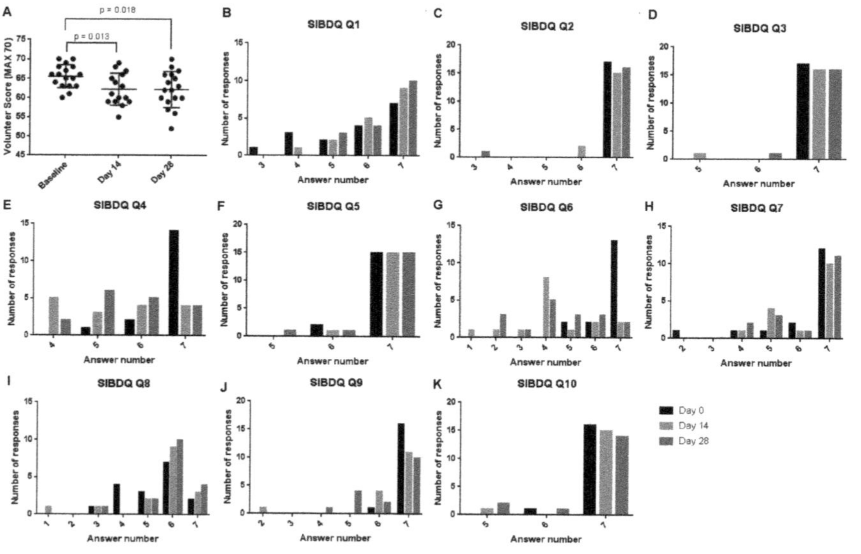

Figure 1. (**A**) S-IBDQ-score obtained from volunteer questions at baseline, day 14 and day 28. Each point represents a volunteer score with the midline and standard deviations plotted. Absolute histograms (**B–K**) of individual patient responses for each of the S-IBDQ questions (1–10 respective) at day 0, 14 and 28. Given p values represent statistical significance using an un-paired t-test with $p < 0.05$ denoting statistical significance.

3.2. Haematological Biochemical Analyses

Haematological and biochemical analyses were assessed over the period of the intervention. There were no significant changes in measured levels of haemoglobin (Figure 2A), creatinine (Figure 2B), Calcium (Figure 2C), alanine aminotransferase (Figure 2D), and magnesium (Figure 2E).

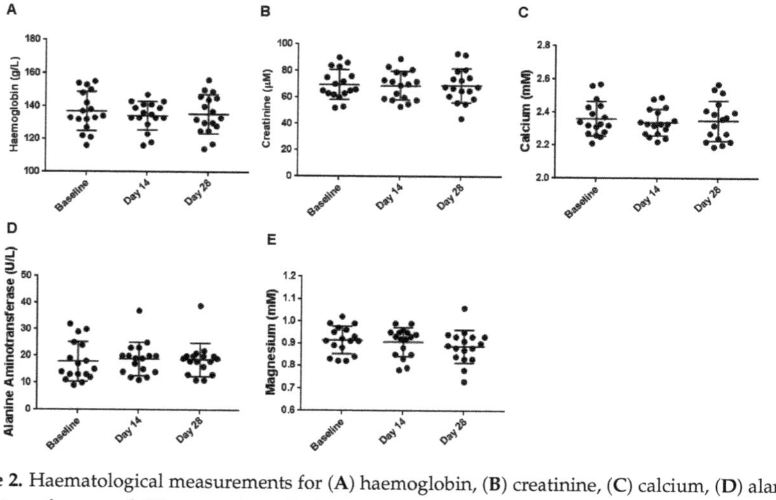

Figure 2. Haematological measurements for (**A**) haemoglobin, (**B**) creatinine, (**C**) calcium, (**D**) alanine aminotransferase and (**E**) magnesium. Each point represents a healthy participant with the midline and standard deviations plotted. 1way ANOVA tests of significance were performed across all time points, and no significance was observed.

3.3. Faecal Iron and Calcium Concentration

Total iron and calcium concentrations from day 0, 14 and 28 remained constant (Figure 3A/C). Average total and free iron concentrations across all samples were 0.37 ± 0.17 and 0.014 ± 0.02 mg/g faecal dry weight, respectively. There were no significant changes in mean free iron values from day 0 to day 14 and 28 (Figure 3B). Average total and free calcium concentrations across all samples were 323.4 ± 380.5 and 36.66 ± 15.15 mg/g faecal dry weight, respectively (Figure 3B/D). No statistical changes in the calcium pools measured were detected (Figure 3D).

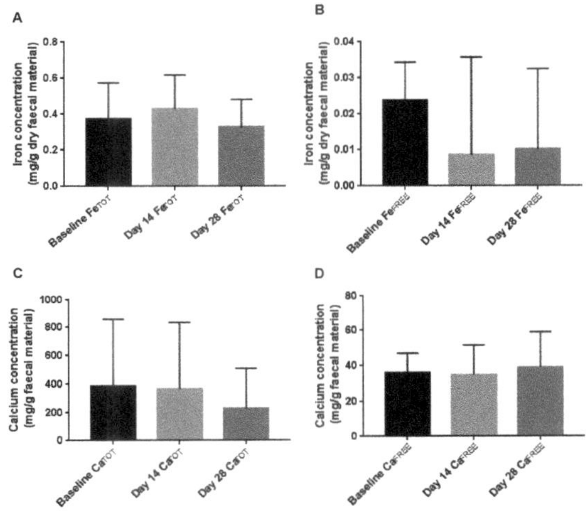

Figure 3. Faecal iron and calcium concentrations (per gram dry weight faecal material) at baseline, day 14 and day 28 of the intervention. (**A**) Total iron, (**B**) free iron, (**C**) total calcium and (**D**) free calcium. Mean values are plotted with error bars denoting standard deviations in the error. No statistical significance (using 1way ANOVA tests) was found between the data sets.

3.4. Faecal Reactive Oxygen Species

Over the time-course of the intervention there were no significant changes in faecal ROS generating potential, with mean MeSH concentrations of 1.8 ± 0.6, 2.23 ± 1.23 and 1.98 ± 0.64 mM/g dry weight faecal material detected at day 0, 14 and 28, respectively (Figure 4).

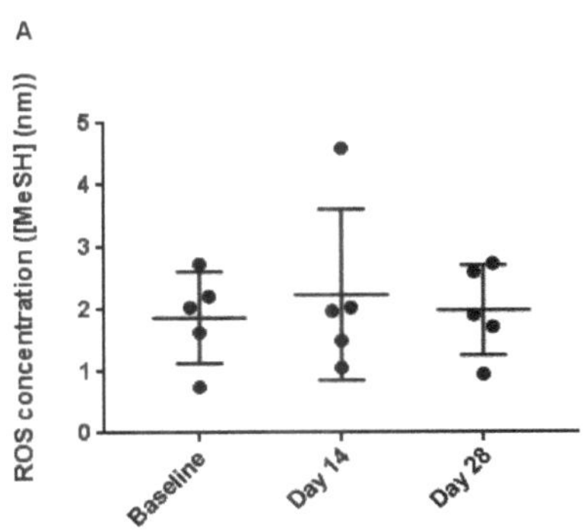

Figure 4. Faecal reactive oxygen species (ROS) concentrations at baseline, day 14 and day 28 of the intervention. No significant changes were observed (1way ANOVA with $p < 0.05$) across all time points.

Correlations between ROS generating potential and concentrations of total/free iron and calcium were also assessed (Figure S4); no significant associations were found. No correlation was found between calcium concentrations and ROS generating potential in any of the pools assessed (Figure S3C/D).

3.5. Healthy Participant Microbiome Analysis

The four major gut phylae showed very similar taxonomic compositions. Although the proportion of Bacterioides increased (Δ21) and Firmicutes (Δ21) reduced by day 28 compared to baseline, this was not significant. No significant changes at the family, genus and species levels were observed between any time points.

Both species richness and diversity were similar for the different sampling time points. There was no significant change in the gut microbial community structures and individual OTUs when compared between any time points.

3.6. Iron, ROS and Ammonia Concentrations within SHIME

To assess in a more controlled environment the impact of MLD on intestinal iron levels an artificial gut model (SHIME) was employed. Total iron changes within the colonic vessels were assessed over the period of the iron/alginate supplementation (day 0, 6 and 15). All iron-fed vessels were established to have an equal and constant iron concentration throughout the period of the study. However, small but significant iron concentration differences were measured for Fe and MLD+Fe vessels at baseline (353 and 400 μM, respectively) and midpoint (395 and 356 μM, respectively) (Figure 5A).

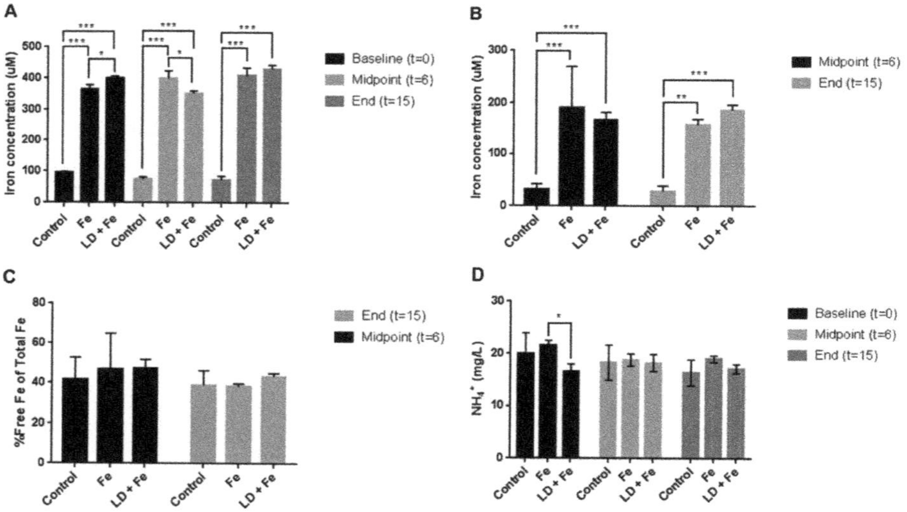

Figure 5. Iron and ammonium parameters within SHIME colonic vessels. Total iron (**A**) was measured across three time points (baseline, midpoint and at the end of the study) in the control, iron (Fe) and iron with MLD (MLD + Fe) vessels (n = 3). Free iron (**B**) was also measured at the midpoint and the end. The percentage of free iron with respect to total Fe was also calculated (**C**). Ammonium concentrations were also measured across three time points (**D**). Bars represent mean values with error bars denoting standard deviations in the error. Statistical significance (using 1way ANOVA tests) is represented with an *, ** or **, where $p < 0.05$, 0.005 and 0.0005 respectively, with n = 3 for each mean value.

Free iron changes were additionally measured (Figure 5B). However, no significant differences were measured between the Fe and MLD+Fe throughout the study period. With total iron concentrations changing within the vessels, the free iron measured was normalised with the total iron, and a percentage of free iron with respect to total iron was calculated (Figure 5C); no significant percentage differences were calculated over the study.

ROS concentrations were also measured throughout the study with no significant changes between the treatment (Fe and MLD+Fe) and control colonic vessels. Additionally, ammonium concentrations for day 0, 6 and 15 of the study are also reported (Figure 5D).

3.7. SHIME SCFA Analysis

Short chain fatty acid (SCFA) assessments were carried out routinely as a measure of bacterial viability throughout the study (Supplementary Figure S5). Relative SCFA concentrations were calculated with respect to total SCFA changes. Overall concentrations of butyric acid concentrations were elevated with MLD+Fe treatment with respect to Fe and control, with statistical significance found at day 6 and day 13 (Figure S5C).

3.8. SHIME Microbiome Analysis

No differences in alpha or beta diversity in the microbial profiles between the three treatment arms and time points were observed. Additionally, there were no significant differences identified in OTUs between the groups.

3.9. Breakdown of Alginate

Alginate breakdown derivatives were assessed using TLC (Figure 6). As a positive control, MLD was degraded by heat, from which degradation products were observed (Figure 6A); increased streaking and higher retention factors (RFs) were observed for longer degradation times. The possibility of degradation of alginate within the colonic vessels of SHIME was also assessed (Figure 6B). SHIME medium from the colonic vessels supplemented with MLD+Fe (at the end of the experiment) demonstrated extensive streaking indicative of degradation, though, of note, the control samples with no alginate similarly demonstrated a streaking profile albeit less pronounced.

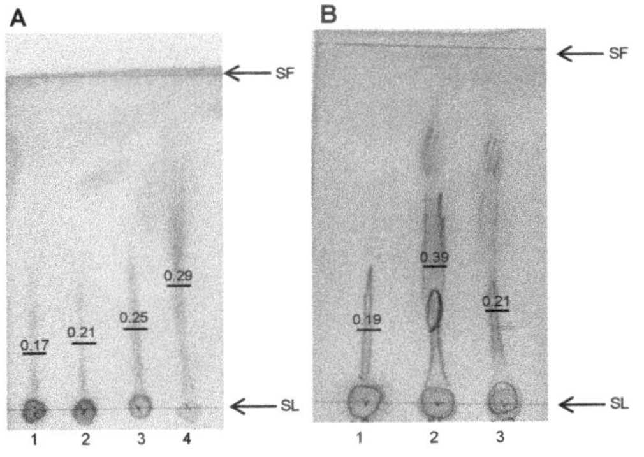

Figure 6. Thin layer chromatography (TLC) plates spotted with (**A**) 1: Native MLD, 2: 5 min heat degraded MLD, 3: 15 min heat degraded MLD, 4: 60 min heat degraded MLD and (**B**) 1: Native MLD, 2: Day 16 MLD colonic vessel and 3: Day 16 control colonic vessel. Solvent Fronts (SF) and Spotting Lines (SL) are indicated. Retention values are calculated and represented on the TLC with a midline.

4. Discussion

Iron excess within the colon has been demonstrated to negatively regulate the microbiome, influencing the development of pathogenic communities [9,11,13,29–31]. The mechanism behind this toxicity is not fully understood, but it is likely due to one of two factors: (1) An abundance of iron promotes the growth of iron-requiring pathogenic bacteria [9] or (2) chemically, it is able to modulate the communities through toxic reactions [6,30,32]. An intervention which would evade both possible mechanisms of microbial dysbiosis would be to chelate the excess iron, rendering it inert, non-reactive and non-utilisable by bacteria. Dietary iron chelators, of which one particular biopolymeric alginate (MLD) was assessed in this first-in-man study, with the primary aim of ascertaining the safety and tolerability associated with consumption. This was assessed through the use of questionnaires and various biomarkers of health including a variety of haematological parameters and changes to the microbiome. In addition, assessments were also undertaken using SHIME to assess the impact of alginate supplementation in a highly controlled intestinal model.

The consumption of alginate in this study was 3.0 g per day, which is greater than that consumed in an average diet (of which there are many estimations, but maximum daily exposure is estimated at 2.1 g daily) [33]. In the healthy participant study, alginate supplementation was well tolerated with little to no side effects observed. However, there were small decreases in the overall S-IBDQ score following alginate consumption. Upon interrogation of the individual questionnaire responses, participants reported an increase in abdominal pain from "none of the time" to "hardly any of the time" and from having problems passing large amounts of gas from "no trouble" to "some trouble". The reason for these changes could be an increase in bacterial fermentation-processes, presumably induced by MLD [34]. This is supported by the observations reported in the artificial gut of a trend for increased butyrate production following alginate supplementation. These physiological changes in the amount of bloating and gas are important to consider in the context of the cohort of patients that such therapies are aimed at, namely those with IBD. As discussed, removal of excess free iron from the bowel of IBD patients would be highly advantageous; however, if the bioactive increased the frequency of the symptoms associated with this disease, the use of such an agent needs to be carefully considered. A limitation of this study was the small cohort size ($n = 17$); the completion of this initial safety assessment of MLD allows for both larger test cohorts and testing in IBD patients in future studies.

To further assess the safety of alginate supplementation, a variety of haematological parameters were assessed. Reassuringly, 28-day consumption of alginate did not reduce the haemoglobin levels, and neither were there any effects on liver biomarkers. In addition, there were no changes in calcium concentration which is important since alginates have a strong affinity for calcium [35]. Since the intestinal microbiome is integral to health and a dysbiosis has been linked with a variety of diseases, alterations in bacterial communities were also assessed following alginate supplementation. Throughout the duration of the study, minimal changes were observed with no net positive or negative growth effects on the bacterial species present. While a shift in the composition of communities present may have been expected with alginate consumption, no such changes were observed. This may be due to the fact that the participants were "free living" with no dietary restrictions. In addition, there inevitably would have been gross differences in diets consumed across all 17 of the participants. Any possible changes induced by alginate consumption may have been overwhelmed by changes influenced by these dietary differences.

These limited changes in the human study are in agreement with the lack of changes observed in the gut model with MLD. Interestingly in this artificial gut model, there were also no alterations in the microbial communities when colonic vessels were co-incubated with a concentration of c. 400 to 600 μM total iron and c. 200 to 300 μM free iron (based on a 70% iron binding capacity by SHIME medium). This is surprising since previous studies have linked iron excess with dysbiosis. The lack of a bacterial dysbiosis in the colonic model could be due to one of several reasons. First, ferrous iron was the only form and species of iron utilised in the SHIME studies, and it could be the case that other forms of iron within the diet cause greater dysbacteriosis. This highlights the importance of understanding the form of iron likely to be present throughout gastrointestinal transit. Second, the SHIME feed contains many iron chelating additives (such as pectin and mucin) [19,36,37] which may both alter the availability of iron to specific bacterial communities and reduce the catalysis of ROS production (and indeed, no increased ROS production was found in iron-supplemented colonic vessels compared to control). The list of chemical contents within the SHIME feed is listed in Supplementary Table S1. Third, the concentration of iron (15 mg/day) may not have been sufficient to induce significant changes. An additional reason as to why limited microbiome changes were found could be due to the iron-species present within the SHIME gut model in comparison to that found in man. Specifically, the results of this study may highlight the importance of dietary ligands on iron and how they impact on the form of iron present upon entering the colon; iron-speciation can radically alter its physico-chemical properties (for example, comparison of the reactivity of a "free" ferrous iron to ferric-oxides) [6,8] and, as such, may alter its activity within a model intestine [38]. The limited changes in ROS concentrations between the iron and control treatments provide evidence to support this, where oxidised and hydrolysed iron forms are unable to partake in redox-reactions. Finally, and most interestingly, is the argument that in the absence of the host compartment (i.e. the colon), iron is unable to mediate any microbiome changes. This may indicate that the observed dysbiosis reported in the literature may be a consequence of alterations in the host which subsequently trigger shifts in the microbiome (in an indirect mechanism). If so, the use of SHIME would only detect a direct action of iron in modulating the microbiome, which was not observed here. This may be a key factor in understanding the mechanism by which iron can induce a microbial dysbiosis within the colon. Such non-direct effects on the microbiome have been reported for other chemicals such as chemotherapeutics [39].

With no changes incident upon iron supplementation within SHIME, a reversal of dysbiotic shifts during co-administration of a fibre-like iron chelator (MLD) would not be observed; this was indeed the case. It is considered that MLD could act as a prebiotic and as such, bacterial changes would be expected, however, no changes were found in this study. This is surprising since the flatulence and bloating which was reported by the healthy volunteers following alginate consumption likely reflects fermentation of the alginate and would, thus, likely result in a change in the microbiota to efficiently utilise this dietary source [40,41]. The degradation experiments and the trend for increased butyrate production, indeed, likely points to fermentation of the alginate. This is further supported by a previous study which reports that alginate is degraded within the colon [40].

Thus, in summary, MLD ingestion was safe and well tolerated, with relatively minor side effects reported of increased flatulence and bloating. Furthermore, there was no evidence of any detrimental changes to the microbiome. This was somewhat surprising, since the reported increased flatulence and bloating likely reflects alginate fermentation and this would likely result in some changes to the composition of the microbiome. To separately assess the impact of iron on the microbiome and whether this could be reversed using an iron chelator, an artificial gut model was utilised, and the results of this were equally surprising with no apparent change in the microbiota with iron and or alginate. This likely reflects the other iron chelating agents found within the food matrix and the need to characterise the exact species of iron in the colon that is responsible for the dysbiotic effects previously reported. It is evident from these studies that any future applications of using alginates as luminal iron chelators require a formulation where bioactivity is targeted to the colon and protected from any dietary agents and/or digestive processes to maximise iron-binding activity.

Supplementary Materials: The following are available online at http://www.mdpi.com/2072-6643/11/3/674/s1; Table S1: M-SHIME nutritional feed components, Figure S1: Short Inflammatory Bowel Disease (IBD) Questionnaire Questions with answer score values, Figure S2: Flow-diagram of healthy volunteer involvement within the study. Patient information sheet (PIS), Figure S3: Visit one (baseline) haematological measurements for (A) haemoglobin, (B) creatinine, (C) alanine aminotransferase, (D) calcium and (E) magnesium. Each point represents a healthy participant, and the mid-line denotes the mean, Figure S4: Correlation plots of both free and total stool iron and calcium concentrations vs. stool ROS concentrations. (A) Total Fe vs. ROS, (B) free Fe vs. ROS, (C) total Ca vs. ROS and (D) free Ca vs. ROS. Dashed-lines represent lines of best fit with associated R2 value and P value representing statistical significance of the observed correlation, Figure S5: Relative short chain fatty acid (SCFA) changes with respect to total SCFA concentrations. Acetic acid (A), propionic acid (B), butyric acid (C) and total branched SCFAs were measured over the time course of the experiment. Each point represents the mean values with error bars denoting standard deviations in the error. Statistical significance (using 1way ANOVA tests, where $p < 0.05$) is represented with an * for Fe + LD vs. Fe alone and $ for Fe + LD vs. Control with n = 3 for each mean value.

Author Contributions: Conceptualization, R.D.H., T.H.I. and C.T.; Formal analysis, R.D.H., M.N.Q. and A.D.B.; Funding acquisition, C.T.; Investigation, R.D.H., P.M. and T.V.d.W.; Methodology, R.D.H. and T.V.d.W.; Project administration, P.M.; Supervision, T.V.d.W., T.H.I. and C.T.; Writing—original draft, R.D.H., T.H.I. and C.T.; Writing—review and editing, T.H.I. and C.T.

Funding: This research was funded by the BBSRC.

Conflicts of Interest: The authors declare no conflict of interest.

References

1. Kortman, G.A.; Raffatellu, M.; Swinkels, D.W.; Tjalsma, H. Nutritional iron turned inside out: Intestinal stress from a gut microbial perspective. *FEMS Microbiol. Rev.* **2014**, *38*, 1202–1234. [CrossRef] [PubMed]
2. Torti, S.V.; Torti, F.M. Iron and cancer: More ore to be mined. *Nat. Rev. Cancer* **2013**, *13*, 342. [CrossRef]
3. Seril, D.N.; Liao, J.; Yang, G.-Y.; Yang, C.S. Oxidative stress and ulcerative colitis-associated carcinogenesis: Studies in humans and animal models. *Carcinogenesis* **2003**, *24*, 353–362. [CrossRef] [PubMed]
4. Werner, T.; Wagner, S.J.; Martínez, I.; Walter, J.; Chang, J.-S.; Clavel, T.; Kisling, S.; Schuemann, K.; Haller, D. Depletion of luminal iron alters the gut microbiota and prevents Crohn's disease-like ileitis. *Gut* **2011**, *60*, 325. [CrossRef] [PubMed]
5. Takagi, T.; Naito, Y.; Mizuhima, K.; Hirai, Y.; Harusato, A.; Okayama, T.; Katada, K.; Kamada, K.; Uchiyama, K.; Handa, O. Heme oxygenase-1 prevents murine intestinal inflammation. *J. Clin. Biochem. Nutr.* **2018**, 17–133. [CrossRef] [PubMed]
6. Carrier, J.; Aghdassi, E.; Platt, I.; Cullen, J.; Allard, J. Effect of oral iron supplementation on oxidative stress and colonic inflammation in rats with induced colitis. *Aliment. Pharmacol. Ther.* **2001**, *15*, 1989–1999. [CrossRef] [PubMed]
7. Glei, M.; Latunde-Dada, G.O.; Klinder, A.; Becker, T.W.; Hermann, U.; Voigt, K.; Pool-Zobel, B.L. Iron-overload induces oxidative DNA damage in the human colon carcinoma cell line HT29 clone 19A. *Mt. Res./Genet. Toxicol. Environ. Mutagen.* **2002**, *519*, 151–161. [CrossRef]
8. Lund, E.K.; Wharf, S.G.; Fairweather-Tait, S.J.; Johnson, I.T. Oral ferrous sulfate supplements increase the free radical–generating capacity of feces from healthy volunteers. *Am. J. Clin. Nutr.* **1999**, *69*, 250–255. [CrossRef] [PubMed]
9. Zimmermann, M.B.; Chassard, C.; Rohner, F.; N'Goran, E.K.; Nindjin, C.; Dostal, A.; Utzinger, J.; Ghattas, H.; Lacroix, C.; Hurrell, R.F. The effects of iron fortification on the gut microbiota in African children: A randomized controlled trial in Côte d'Ivoire. *Am. J. Clin. Nutr.* **2010**, *92*, 1406–1415. [CrossRef]
10. Fang, S.; Zhuo, Z.; Yu, X.; Wang, H.; Feng, J. Oral administration of liquid iron preparation containing excess iron induces intestine and liver injury, impairs intestinal barrier function and alters the gut microbiota in rats. *J. Trace Elem. Med. Biol.* **2018**, *47*, 12–20. [CrossRef] [PubMed]
11. Paganini, D.; Zimmermann, M.B. The effects of iron fortification and supplementation on the gut microbiome and diarrhea in infants and children: A review. *Am. J. Clin. Nutr.* **2017**, *106*, 1688S–1693S. [CrossRef]
12. Constante, M.; Fragoso, G.; Calvé, A.; Samba-Mondonga, M.; Santos, M.M. Dietary heme induces gut dysbiosis, aggravates colitis, and potentiates the development of adenomas in mice. *Front. Microbiol.* **2017**, *8*, 1809. [CrossRef]
13. Jaeggi, T.; Kortman, G.A.; Moretti, D.; Chassard, C.; Holding, P.; Dostal, A.; Boekhorst, J.; Timmerman, H.M.; Swinkels, D.W.; Tjalsma, H. Iron fortification adversely affects the gut microbiome, increases pathogen abundance and induces intestinal inflammation in Kenyan infants. *Gut* **2015**, *64*, 731–742. [CrossRef] [PubMed]
14. Chua, A.C.; Klopcic, B.R.; Ho, D.S.; Fu, S.K.; Forrest, C.H.; Croft, K.D.; Olynyk, J.K.; Lawrance, I.C.; Trinder, D. Dietary iron enhances colonic inflammation and IL-6/IL-11-Stat3 signaling promoting colonic tumor development in mice. *PLoS ONE* **2013**, *8*, e78850. [CrossRef] [PubMed]
15. Radulescu, S.; Brookes, M.J.; Salgueiro, P.; Ridgway, R.A.; McGhee, E.; Anderson, K.; Ford, S.J.; Stones, D.H.; Iqbal, T.H.; Tselepis, C. Luminal iron levels govern intestinal tumorigenesis after Apc loss in vivo. *Cell Rep.* **2012**, *2*, 270–282. [CrossRef] [PubMed]
16. Collings, R.; Harvey, L.J.; Hooper, L.; Hurst, R.; Brown, T.J.; Ansett, J.; King, M.; Fairweather-Tait, S.J. The absorption of iron from whole diets: A systematic review. *Am. J. Clin. Nutr.* **2013**, *98*, 65–81. [CrossRef] [PubMed]

17. Powell, J.J.; Cook, W.B.; Hutchinson, C.; Tolkien, Z.; Chatfield, M.; Pereira, D.I.; Lomer, M.C. Dietary fortificant iron intake is negatively associated with quality of life in patients with mildly active inflammatory bowel disease. *Nutr. Metab.* **2013**, *10*, 9. [CrossRef] [PubMed]
18. Kaitha, S.; Bashir, M.; Ali, T. Iron deficiency anemia in inflammatory bowel disease. *World J. Gastrointest. Pathophysiol.* **2015**, *6*, 62–72. [CrossRef] [PubMed]
19. Debon, S.J.; Tester, R.F. In vitro binding of calcium, iron and zinc by non-starch polysaccharides. *Food Chem.* **2001**, *73*, 401–410. [CrossRef]
20. Berner, L.A.; Hood, L.F. Iron binding by sodium alginate. *J. Food Sci.* **1983**, *48*, 755–758. [CrossRef]
21. Wölbling, R.; Becker, G.; Forth, W. Inhibition of the intestinal absorption of iron by sodium alginate and guar gum in rats. *Digestion* **1980**, *20*, 403–409. [CrossRef] [PubMed]
22. Lee, K.Y.; Mooney, D.J. Alginate: Properties and biomedical applications. *Prog. Polym. Sci.* **2012**, *37*, 106–126. [CrossRef] [PubMed]
23. Hampson, F.; Farndale, A.; Strugala, V.; Sykes, J.; Jolliffe, I.; Dettmar, P. Alginate rafts and their characterisation. *Int. J. Pharm.* **2005**, *294*, 137–147. [CrossRef]
24. Horniblow, R.D.; Latunde-Dada, G.O.; Harding, S.E.; Schneider, M.; Almutairi, F.M.; Sahni, M.; Bhatti, A.; Ludwig, C.; Norton, I.T.; Iqbal, T.H. The chelation of colonic luminal iron by a unique sodium alginate for the improvement of gastrointestinal health. *Mol. Nutr. Food Res.* **2016**, *60*, 2098–2108. [CrossRef]
25. Wawer, A.A.; Harvey, L.J.; Dainty, J.R.; Perez-Moral, N.; Sharp, P.; Fairweather-Tait, S.J. Alginate inhibits iron absorption from ferrous gluconate in a randomized controlled trial and reduces iron uptake into caco-2 cells. *PLoS ONE* **2014**, *9*, e112144. [CrossRef]
26. Van den Abbeele, P.; Roos, S.; Eeckhaut, V.; MacKenzie, D.A.; Derde, M.; Verstraete, W.; Marzorati, M.; Possemiers, S.; Vanhoecke, B.; Van Immerseel, F. Incorporating a mucosal environment in a dynamic gut model results in a more representative colonization by lactobacilli. *Microb. Biotechnol.* **2012**, *5*, 106–115. [CrossRef] [PubMed]
27. Giuliani, C.; Marzorati, M.; Innocenti, M.; Vilchez-Vargas, R.; Vital, M.; Pieper, D.H.; Van de Wiele, T.; Mulinacci, N. Dietary supplement based on stilbenes: A focus on gut microbial metabolism by the in vitro simulator M-SHIME®. *Food Funct.* **2016**, *7*, 4564–4575. [CrossRef]
28. Li, M.; Li, G.; Shang, Q.; Chen, X.; Liu, W.; Pi, X.; Zhu, L.; Yin, Y.; Yu, G.; Wang, X. In vitro fermentation of alginate and its derivatives by human gut microbiota. *Anaerobe* **2016**, *39*, 19–25. [CrossRef]
29. Lee, T.; Clavel, T.; Smirnov, K.; Schmidt, A.; Lagkouvardos, I.; Walker, A.; Lucio, M.; Michalke, B.; Schmitt-Kopplin, P.; Fedorak, R. Oral versus intravenous iron replacement therapy distinctly alters the gut microbiota and metabolome in patients with IBD. *Gut* **2017**, *66*, 863–871. [CrossRef] [PubMed]
30. Mahalhal, A.; Williams, J.M.; Johnson, S.; Ellaby, N.; Duckworth, C.A.; Burkitt, M.D.; Liu, X.; Hold, G.L.; Campbell, B.J.; Pritchard, D.M. Oral iron exacerbates colitis and influences the intestinal microbiome. *PLoS ONE* **2018**, *13*, e0202460. [CrossRef] [PubMed]
31. Tang, M.; Frank, D.N.; Hendricks, A.E.; Ir, D.; Esamai, F.; Liechty, E.; Hambidge, K.M.; Krebs, N.F. Iron in micronutrient powder promotes an unfavorable gut microbiota in Kenyan infants. *Nutrients* **2017**, *9*, 776. [CrossRef] [PubMed]
32. Lobo, V.; Patil, A.; Phatak, A.; Chandra, N. Free radicals, antioxidants and functional foods: Impact on human health. *Pharm. Rev.* **2010**, *4*, 118. [CrossRef] [PubMed]
33. Additives, E.P.O.F.; Food, N.S.A.t.; Younes, M.; Aggett, P.; Aguilar, F.; Crebelli, R.; Filipič, M.; Frutos, M.J.; Galtier, P.; Gott, D.; et al. Re-evaluation of alginic acid and its sodium, potassium, ammonium and calcium salts (E 400–E 404) as food additives. *EFSA J.* **2017**, *15*, e05049.
34. Cummings, J.H.; Macfarlane, G.T. Gastrointestinal effects of prebiotics. *Br. J. Nutr.* **2002**, *87*, S145–S151. [CrossRef] [PubMed]
35. Coudray, C.; Bellanger, J.; Castiglia-Delavaud, C.; Remesy, C.; Vermorel, M.; Rayssignuier, Y. Effect of soluble or partly soluble dietary fibres supplementation on absorption and balance of calcium, magnesium, iron and zinc in healthy young men. *Eur. J. Clin. Nutr.* **1997**, *51*, 375. [CrossRef]
36. PLAIT, S.R.; Clydesdale, F.M. Mineral binding characteristics of lignin, guar gum, cellulose, pectin and neutral detergent fiber under simulated duodenal pH conditions. *J. Food Sci.* **1987**, *52*, 1414–1419. [CrossRef]
37. Powell, J.; Jugdaohsingh, R.; Thompson, R. The regulation of mineral absorption in the gastrointestinal tract. *Proc. Nutr. Soc.* **1999**, *58*, 147–153. [CrossRef]

38. Pereira, D.I.; Bruggraber, S.F.; Faria, N.; Poots, L.K.; Tagmount, M.A.; Aslam, M.F.; Frazer, D.M.; Vulpe, C.D.; Anderson, G.J.; Powell, J.J. Nanoparticulate iron (III) oxo-hydroxide delivers safe iron that is well absorbed and utilised in humans. *Nanomed. Nanotechnol. Biol. Med.* **2014**, *10*, 1877–1886. [CrossRef]
39. Vanlancker, E.; Vanhoecke, B.; Stringer, A.; Van de Wiele, T. 5-Fluorouracil and irinotecan (SN-38) have limited impact on colon microbial functionality and composition in vitro. *PeerJ* **2017**, *5*, e4017. [CrossRef]
40. Wang, Y.; Han, F.; Hu, B.; Li, J.; Yu, W. In vivo prebiotic properties of alginate oligosaccharides prepared through enzymatic hydrolysis of alginate. *Nutr. Res.* **2006**, *26*, 597–603. [CrossRef]
41. Chen, X.; Sun, Y.; Hu, L.; Liu, S.; Yu, H.; Li, R.; Wang, X.; Li, P. In vitro prebiotic effects of seaweed polysaccharides. *J. Oceanol. Limnol.* **2018**, *36*, 926–932. [CrossRef]

nutrients

MDPI

Article

Iron (II) Citrate Complex as a Food Supplement: Synthesis, Characterization and Complex Stability

Daniele Naviglio [1,†], Maria Michela Salvatore [1,†], Marianna Limatola [1], Ciro Langella [2], Stefano Faralli [3], Martina Ciaravolo [1], Anna Andolfi [1], Francesco Salvatore [1] and Monica Gallo [2,*]

1 Department of Chemical Sciences, University of Naples Federico II, via Cintia, 4, 80126 Naples, Italy; naviglio@unina.it (D.N.); mariamichela.salvatore@unina.it (M.M.S.); limatolamarianna@gmail.com(M.L.); martinaciaravolo@gmail.com (M.C.); andolfi@unina.it (A.A.); frsalvat@unina.it (F.S.)

2 Department of Molecular Medicine and Medical Biotechnology, University of Naples Federico II, via Pansini, 5, 80131 Naples, Italy; cirolangella88@gmail.com

3 Medical Center, Piazzale Luigi Cadorna, 9, 20123 Milan, Italy; info@faralli.net

* Correspondence: mongallo@unina.it; Tel.: +39-081-746-3117

† These authors contributed equally to this work.

Received: 15 September 2018; Accepted: 29 October 2018; Published: 3 November 2018

Abstract: Iron deficiency represents a widespread problem for a large part of the population, especially for women, and has received increasing attention in food/supplement research. The contraindications of the iron supplements commercially available (e.g., imbalances in the levels of other essential nutrients, low bioavailability, etc.) led us to search for a possible alternative. In the present work, a rapid and easy method to synthetize a solid iron (II) citrate complex from iron filings and citric acid was developed to serve, eventually, as a food supplement or additive. In order to state its atomic composition and purity, an assortment of analytical techniques was employed (e.g., combustion analysis, thermogravimetry, X-ray diffractometry, UV/Vis spectrophotometry, etc.). Results demonstrate that the synthesized crystalline solid corresponds to the formula $FeC_6H_6O_7 \cdot H_2O$ and, by consequence, contains exclusively iron (II), which is an advantage with respect to existing commercial products, because iron (II) is better absorbed than iron (III) (high bioavailability of iron).

Keywords: iron citrate (II); food supplement; nutraceuticals; iron deficiency; anaemia

1. Introduction

Iron is a trace element necessary for the human body because it enters the constitution of haemoglobin, myoglobin and different iron-dependent enzymes [1]. In addition to erythropoiesis, iron is essential for mitochondrial function, DNA synthesis and repair, and many enzymatic reactions required for cell survival. Iron deficiency (generally indicated as anaemia) may contribute to cognitive developmental defects in children, poor physical performance, and unfavourable pregnancy outcomes [2–6]. Dysregulation of iron metabolism contributes to various human pathologies, including iron overload diseases and cancer. Great strides in understanding iron metabolism have been made in the past several years, including the identification of new essential proteins in human iron metabolism and novel roles for iron in both normal cellular processes and in pathologies such as cancer [7]. In addition, iron is deeply linked to cell death; traditionally, iron is thought to contribute to cell death pathways through reactive oxygen species (ROS) production. Iron and ROS are increasingly recognized as important initiators and mediators of cell death in a variety of organisms and pathological situations. Recently, an ROS-independent role for iron in modulating apoptosis (programmed cell death) was defined. Furthermore, an entirely new mode of iron-dependent cell death, termed ferroptosis, was described [8]. The absorption of iron essentially depends on three main factors: (1) the iron content of the diet; (2) the liberation of iron contained in food by the digestive processes and its availability to be taken from the mucosa of the small intestine; (3) intestinal conditions at the intraluminal level that greatly influence the final availability of iron. Moreover, there are two forms of iron: iron-heme and non-heme iron (the latter is also called inorganic iron), because their bioavailability is very different and greatly affects the performanceof the potential absorption of iron in the diet [9].

Heme iron is found only in foods of animal origin, particularly in meat, as it is present in muscle haemoproteins (dairy products, on the contrary, are completely devoid of them), and also in haemoproteins of fish and eggs, and it is absorbed better than non-heme iron from foods. Its absorption, which is around 25% (w/w), is independent of the composition of the diet because it is absorbed intact as a porphyrin complex without interference from the other constituents of the diet [10]. The absorption of non-heme iron, on the contrary, depends on the composition of the diet and the presence of enhancers and inhibitors of absorption [11] as well as the oxidation state of the metal [12]. Non-heme iron is found in cereals and vegetables and its absorption is generally low; only about 2% to 20% of the iron available from vegetable sources is absorbed. The absorption of the iron by the organism takes place mainly in the duodenum and in the proximal section of the jejunum, and transited through the enterocyte to release the free iron into blood circulation. Therefore, ferrous iron, being much more soluble than ferric iron (at the same pH value), can be more easily absorbed, providing a greater opportunity for chelation and a higher exposure of the mucosa to solubilized forms before precipitation occurs [13]. The chelate forms of ferric iron that remain in solution can probably be taken from the mucosal cells; in any case, the iron made free is transferred to the plasma in a form that can be linked to the transferrin. Ferric iron is absorbed only after its reduction to ferrous iron; this reduction process can occur in the stomach [14]. The indications given for dietary iron intake are based on the losses and quantities necessary for growth and to build up reserves but taking into account the bioavailability of this nutrient in the diet itself. Iron deficiency is responsible for anaemia, a widespread disease in Europe [15]. Clinically, iron deficiency anaemia occurs with asthenia, pallor, tachypnoea and tachycardia [16]. Anaemia should be identified on the basis of haemoglobin values: values between 13 and 16 g/dL in men and between 12 and 16 g/dL, in women, are considered an expression of normality [17]. During pregnancy and/or menstruation, the values are lower due to haemodilution or blood losses. When the need for iron in the body is not satisfied, it is advisable to resort to supplements to treat or prevent iron deficiency or simply to compensate for its deficiency. Dietary supplements are widely used and offer the potential to improve health if appropriately targeted to those in need. Inadequate nutrition and micronutrient deficiencies are prevalent conditions that adversely affect global health. Although improvements in diet quality are essential to address these issues, dietary supplements or food fortification could help individuals at risk of deficiencies meet the requirements [18–20]. It is noteworthy that iron supplements can also cause side effects or imbalances in the levels of other essential nutrients, especially metals such as copper and zinc [21]. These contraindications can lead the individual to intolerance of the iron supplement, or even to the possible rejection of it. It is therefore necessary to evaluate the quality of the supplement. For this reason, there are many iron formulations alternatives to prevent or treat iron deficiency anaemia. Organic formulations such as iron gluconate, iron pyrophosphate, iron bis-glycinate, etc. have components that reduce side effects and improve solubility. However, it was found that these formulations did not always have an immediate effect on the individual and were therefore not often used to manage severe iron deficiency [22].

In the past, to compensate for iron deficiency, a popular method involved the use of an apple "contaminated" with iron. The method consisted of inserting iron nails into an apple for a night, in order to transfer a small part of the iron to the apple, because organic acids naturally contained in the apples (alpha-hydroxy acids, such as malic acid, tartaric acid, citric acid etc.) made the nails rust and the iron thus yielded was absorbed by the pulp of the fruit in the form of ferrous organic complexes (among which is the citrate iron (II) complex). The following day, the apple with the composite extracted iron from the nail specimen was ingested by the patient. In the literature, there is a patent on the preparation of iron citrate [23], which has never been commercialized due to the high costs and complex procedures of the preparation of the compound. The hydrothermal synthesis of coordination polymer of iron (II) citrate was described by Birsa Čelič et al. [24] as potential candidates for biomedical applications. Starting from these premises, the current study evaluated the production and characterization of an iron (II) citrate complex by direct reaction between iron filings and citric acid in aqueous solution.

2. Materials and Methods

2.1. General Experimental Information

All reagents, solvents and standards used to produce the iron citrate neutral complex were obtained from different suppliers and all were of high purity: iron filings (>99%), citric acid monohydrate (>99.5%), iron sulphate heptahydrate, 1,10-ortho-phenanthroline and potassium thiocyanate pure for analysis were purchased from Sigma-Aldrich (Buchs, Switzerland). Ultrapure water was produced by a milli-Q generator (Millipore, Bedford, MA, USA). Potassium sulfocyanide pure for analysis from Carlo Erba (Milan, Italy). Analytical balance series B 204 S (Mettler-Toledo, Germany), freeze dryer mod. Heto Lyolab 3000 (Analytical De Mori, Milan, Italy). UV-Vis spectrophotometer, mod UV-1601 (Shimadzu, Tokyo, Japan), quartz cuvettes with optical path of 1.00 cm. Automatic powder diffractometer Panalytical Empyrean Powder (Santa Barbara, CA, USA) and automatic analyser for the determination of total organic carbon Skalar Analytical B.V (Breda, The Netherlands) were used. Thermogravimetric analyser TGA400 Perkin Elmer (Norwalk, CT, USA).

2.2. Synthesis of Iron Citrate Complex

Citric acid monohydrate (21 g) were added to 500 mL of ultrapure water and the solution was heated under magnetic stirring. The iron filings (5.5 g) were added when the temperature of the solution reached 90 °C and the mixture was cooled at room temperature after the complete reaction of the iron. The precipitate grey/pearly salt was washed with water and filtered off under a vacuum with a paper filter; finally, the residue was freeze-dried after freezing at −20 °C.

2.3. Thermogravimetric Analysis (TGA)

Thermogravimetric analysis (TGA) was conducted for iron citrate (~100 mg) in a temperature range from 30 °C to 900 °C. Data are shown as the average of three independent replicates.

2.4. Determination of Iron Oxidation State

The colorimetric method based on the complex ion formed with three molecules of 1,10-phenanthroline and one of iron (II) ion was used to detect the oxidation state of iron (II) ion [25].

The complexation of the iron (III) ion with thiocyanate (SCN^-) has been useful to detect the oxidation state of iron (III) ion as reported by Lister et al. [26]. In this case, the sensitivity of these colorimetric methods (0.1–0.5 ppm) was sufficient for our purposes.

The eventual colour (red) appearing in the solution, during the first or the second mentioned assays, indicates the presence of iron (II) or iron (III) ions, respectively.

Spectrophotometric analyses were performed, for quantitative analyses, at the wavelength of 520 nm and 480 nm, that are maximum absorption values of 1,10-phenanthroline and thiocyanate complexes, respectively.

For the spectrophotometric assay of iron, the required solutions were prepared from analytical grade ammonium iron (II) sulphate [$(NH_4)_2Fe(SO_4)_2 \times 6H_2O$], 1,10-phenanthroline, potassium thiocyanate (KSCN), hydroxylamine hydrochloride ($NH_2OH \times HCl$), sodium acetate (NaAc) and 99.999% H_2SO_4 (Aldrich product # 339741).

2.5. Complex Stability

Iron citrate (50 mg) was added to 100 mL of ultrapure water. The pH value (6.5) of the solution was modified using NaOH (0.1 M) or HCl (0.1 M). The evaluated pH range was from 2 to 12. The stability of the complex was evaluated via the spectrophotometric techniques described in Section 2.4.

2.6. X-ray Diffractometry

The X-ray diffraction profiles of the synthesized samples were recorded with an automatic powder diffractometer, using the Kα radiation of Cu filtered with Ni, which corresponds to a wavelength of 1.5418 Å. The profiles were recorded using a continuous scan in the range of the 2θ diffraction angle between 5 and 60°.

2.7. Attenuated Total Reflectance (ATR) Analysis

A weighted amount of iron citrate was set in a plate of zinc selenide (ZnSe) and the depth of penetration in the sample of the incident ray was about 0.5 µm. The spectra were acquired between 4000 cm^{-1} and 700 cm^{-1} and reported in absorbance units, with the last spectrum recorded during cleaning serving as the reference. Where needed, signals from the water gas-phase spectrum in the 4000–700 cm^{-1} range were subtracted.

2.8. Total Organic Carbon Analysis (TOC)

Iron (II) citrate (70 mg) in solid form was subjected to elemental carbon analysis in order to evaluate the percentage of carbon present in it. Samples were oxidized with phosphoric acid and ammonium persulphate in the reaction chamber and then gaseous flow was divided into two parts in order to test the amount of organic carbon and inorganic carbon. Hence, one of them was used to evaluate the inorganic carbon obtained after oxidation via the thermal combustion method. The second part was used to test the organic carbon and was bombarded with UV ray emitted by a mercury vapor lamp. In this way, free radicals react with carbon to produce CO_2. Both sample parts were passed to the membranes to measure their conductivity via an automatic analyser for the determination of total organic carbon.

3. Results and Discussion

The iron (II) citrate complex was obtained by direct reaction from the reactants as excess citric acid and iron filings, exploiting a hot oxidation reaction. The synthesis reactions were carried out by maintaining an excess of citric acid in the reaction environment in such a way as to guarantee the complete transformation of the introduced iron. The iron has been used in the form of filings in order to have a higher reaction surface for the iron itself. As reported in Section 2, when the solution of citric acid in water was almost boiling, the iron filings were added. The redox reaction of iron with citric acid in the presence of water led to the formation of the iron citrate with a stoichiometric 1:1 reaction, in which Fe was coordinated with two oxygen atoms, then releasing hydrogen in the gaseous form. The slow in situ generation of iron (II) favours the formation of a crystalline solid keeping oversaturation to low levels. Furthermore, the reducing environment avoided the formation of Fe (III).

The reaction is as follows:

$$\mathrm{Fe\ (s) + C_6H_8O_7\ (s) = FeC_6H_6O_7 + H_2\ (g).} \quad (1)$$

Based on the stoichiometric ratio 1:1 of the reagents (citric acid and iron filings), the iron theoretical percentage is 22.7% $FeC_6H_6O_7$ and 21.2% ($FeC_6H_6O_7 \cdot H_2O$).

$$\%\mathrm{Fe}(\mathrm{FeC_6H_6O_7}) = \frac{\mathrm{AW(Fe)}}{\mathrm{MW(FeC_6H_6O_7)}} \times 100 = \frac{55.845}{245.95} \times 100 = 22.7\% \quad (2)$$

$$\%Fe(FeC_6H_6O_7{\cdot}H_2O) = \frac{AW(Fe)}{MW(FeC_6H_6O_7{\cdot}H2O)} \times 100 = \frac{55.845}{263.963} \times 100 = 21.2\% \quad (3)$$

Thermogravimetry led to the easy determination of the percentage by weight of iron in the iron citrate powder. In fact, from the data obtained, the percentage of iron in the samples was 21.1% ± 0.4 (w/w). Hence, the results are in good agreement with the theoretical value calculated for ($FeC_6H_6O_7{\cdot}H_2O$) (Figure 1).

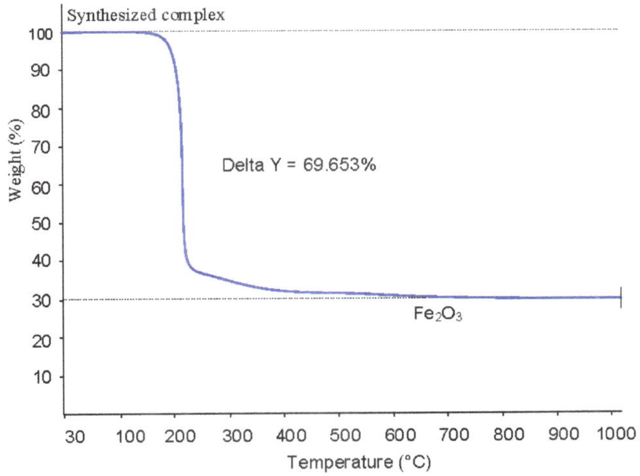

Figure 1. TGA analysis of iron citrate powder. The temperature increased from 30 °C to 900 °C at 200 °C min^{-1}.

Total carbon analysis of the solid iron citrate complex was also performed. Based on the stoichiometric ratio 1:1 of the components, the carbon theoretical percentage is 29.3% ($FeC_6H_6O_7$) and 27.3% ($FeC_6H_6O_7{\cdot}H_2O$).

$$\%C(FeC_6H_6O_7) = \frac{AW(C) \times 6}{MW(FeC_6H_6O_7)} \times 100 = \frac{72.066}{245.95} \times 100 = 29.3\% \quad (4)$$

$$\%C(FeC_6H_6O_7{\cdot}H_2O) = \frac{AW(C) \times 6}{MW(FeC_6H_6O_7{\cdot}H2O)} \times 100 = \frac{72.066}{263.963} \times 100 = 27.3\% \quad (5)$$

Collected data are in good agreement with the theoretical value for $FeC_6H_6O_7{\cdot}H_2O$. In fact, the total carbon percentage in the iron citrate samples is 27.2 ± 0.1% (w/w) (average of three samples).

In light of the total carbon analysis, the thermogravimetric data in Figure 1 can be interpreted in an alternative way to prove that the content of hydrogen and oxygen of the iron citrate complex is consistent with the elemental composition $FeC_6H_6O_7{\cdot}H_2O = FeC_6H_8O_8$.

In fact, since during the TGA analysis, the iron citrate complex is converted to Fe_2O_3(s), the weight loss of 69.653% exposed in Figure 1 implies that 69.653 g of carbon, hydrogen and oxygen are lost per 100 g of the iron complex.

Now, if the atomic composition of the iron complex, with respect to C, H and O, is $C_xH_yO_w$, the 69.653 g per 100 g of the complex that is lost during the TGA analysis corresponds to the elemental composition $C_xH_yO_{w-1.5}$ (because not all oxygen is lost and 1.5 atoms of oxygen per mole of the iron complex remain in the Fe_2O_3 (s) product).

After the carbon analysis, we know that 27.2 ± 0.1 g of carbon are obtained per 100 g of the solid iron citrate complex. In consequence, the amount of ($H_yO_{w-1.5}$) from the array of atoms

$C_xH_yO_{w-1.5}$ that is lost per 100 g of iron citrate complex is experimentally determined to be $(H_yO_{w-1.5}) = 69.653-27.2 = 42.5 \pm 0.1$ g.

If we assume that the solid iron citrate complex is $FeC_6H_6O_7$, then $(H_yO_{w-1.5}) = (H_6O_{5.5})$. Under this assumption, only 38.2 g of $(H_6O_{5.5})$ would be obtained from 100 g of $FeC_6H_6O_7$.

On the contrary, if we assume that the solid iron complex is $FeC_6H_6O_7{\cdot}H_2O = FeC_6H_8O_8$, then $(H_yO_{w-1.5}) = (H_8O_{6.5})$ and, from 100 g of $FeC_6H_6O_7{\cdot}H_2O$, would theoretically be obtained 42.43 g of (H_8O_6). This theoretical value (42.43 g) agrees very well with the value found by combining the TGA and total carbon analysis (42.5 ± 0.1 g) and confirms the composition $FeC_6H_6O_7{\cdot}H_2O$ for the iron citrate complex (as would the elemental analysis of hydrogen and oxygen in the solid).

The spectrophotometric method, based on the complex formed by iron (II) with 1,10-phenanthroline (Phen), i.e., $Fe^{2+} + 3Phen \rightleftharpoons Fe(Phen)_3{}^{2+}$, has been employed both for the analysis of total iron and iron (II) amounts in the iron citrate complex, using two slightly different procedures that are described below.

A 1g/L iron (II) standard solution was prepared by dissolving 7.0226 g of ultrapure $(NH_4)_2Fe(SO_4)_2{\times}6H_2O$ (MW = 392.14 u) in deaerated water up to the volume of a 1L volumetric flask. From this, a secondary standard, 100 mg/L of iron (II), was prepared by diluting 1:10 with deaerated water.

Ten calibrating solutions, of known iron (II) concentration, ranging from 0.2 to 2.0 mg/L (or, $3.58{\times}10^{-5}$ M to $3.58{\times}10^{-6}$ M), were prepared by mixing, in 100-mL volumetric flasks, measured volumes of the 100 mg/L Fe(II) standard solution with fixed volumes of 2 M H_2SO_4, 10% *w/w* $NH_2OH{\times}HCl$, 0.25% *w/w* 1, 10-phenanthroline (Phen) and 2.0 M sodium acetate, according to Table 1.

Table 1. Preparation of iron (II) calibrating solutions.

Standard Fe (II) 100 mg/L	H_2SO_4 2M	$NH_2OH \times HCl$ 10% *w/w*	Phen 0.25% *w/w*	NaAc 2 M	Total Volume (with H_2O)	Conc. Fe (II)
mL	mL	mL	mL	mL	mL	mg/L
0.200–2.00	2.00	1.00	5.00	5.00	100	0.200–2.000

The pH of the calibrating solutions was 4.0 ± 0.2. Addition of an excess of hydroxylamine hydrochloride, which reduces iron (III) to iron (II) ($2NH_2OH + 4Fe^{3+} \rightarrow N_2O + 4Fe^{2+} + H_2O + 4H^+$), keeps iron completely in the +2 oxidation state.

The spectrophotometer was calibrated by measuring the absorbance at 520 nm of the calibration solutions in Table 1, using 1.00 cm quartz cuvettes. Absorbance was measured against a reagent blank that was prepared in the same way as the calibration solutions, except that the 100 mg/L standard solution was not added.

The experimental data points (Abs., conc.) were well interpolated by a least squares regression line with a coefficient of determination $R^2 = 0.9993$ (Abs. = Absorbance at 520 nM; Conc. = iron(II) concentration, mg/L).

Then, to the experimental (Abs., conc.) data, the MS Excel statistical function LINEST was applied, which is a matrix function that calculates several statistical parameters from an array of (y, x) data points.

From collected (Abs., conc.) data, the following statistics were evaluated: the slope ($b = 0.191$ L/mg) and intercept on the absorbance axis ($a = 0.0097$) of the least square regression line ($y = a + bx$) through the experimental points; the standard deviations of slope ($s_b = 0.0017$ L/mg) and of intercept ($s_a = 0.0021$); and, finally, the standard deviation of experimental points around the regression line ($s_{y/x} = 0.0030$).

From the standard deviation of intercept (s_a) and the slope (b) of the regression line, the limit of detection (LOD) of the method is evaluated to be 0.036 mg/L (i.e., LOD = $3.3 \times s_a/b$ mg/L) [27].

The standard deviation around the regression line ($s_{y/x}$) is used to evaluate the standard deviation of iron concentrations by interpolating the calibration curve with the measured absorbance of samples.

Iron, in the iron citrate complex, was determined by weighing 0.2500 ± 0.0002 g of the solid sample directly in a 250 mL volumetric flask. The solid sample was then completely dissolved by adding 2 M H_2SO_4 up to the volume of the volumetric flask. Before use, a stream of nitrogen was passed through the 2 M H_2SO_4 solution for several hours in order to expel the oxygen, which would oxidize iron (II) to iron (III). After preparation, much care was taken to avoid oxygen absorption by the solution and to preserve the original oxidation state of iron in the dissolved solid.

For this solution, the total iron concentration was determined very simply, by transferring 0.5 mL of the solution to a 100-mL volumetric flask and adding to it the same volume of reagents used during the preparation of the calibrating solutions (see Table 1). However, to keep the pH of the measured sample solution within the range of the calibration standards (4 ± 0.2), the volume of 2 M H_2SO_4 solution added was reduced to 1.5 mL.

These operations result in the determination, from the measured absorbance, of the total iron concentration in the above iron citrate complex solution. In fact, even if iron (III) is present in the sample, it will be reduced to iron (II) by hydroxylamine.

This experiment was repeated three times and the three values of the total iron concentration were averaged. To the average value of total iron concentration, a 95% confidence level interval was attached, as evaluated from the standard deviation around the regression ($s_{y/x}$) [27].

The total iron concentration in the sample solution, evaluated assuming for the iron citrate complex the composition $FeC_6H_6O_7 \cdot H_2O$, was 1.058 ± 0.010 mg/L, and compares very well with the average concentration, 1.082 ± 0.025 mg/L, determined from the three replicates of the analysis.

Exactly the same experiment was repeated with a second 0.5-mL portion of the iron citrate complex solution, except that, in this case, the hydroxylamine solution was substituted with a sodium fluoride (NaF) solution, which masks iron (III). Furthermore, before use, oxygen was expelled from all the added solutions by purging with a nitrogen flux and the prepared solution, carefully kept out of contact with the atmosphere, was, after a very short delay, presented to the spectrophotometer and its absorbance measured. In this way, only iron (II) was determined from the measured absorbance.

As in the previous case, this experiment was repeated three times and three values of the iron (II) concentration were obtained and averaged.

Obviously, if the dissolved iron citrate complex contains a significant fraction of iron in the form of iron (III), the total iron concentration will be greater than the iron (II) concentration and from the difference (total iron concentration – iron (II) concentration), it is possible to evaluate the amount of iron present as iron (III).

However, the average of the three iron (II) concentrations, determined by the above procedure, was 1.070 ± 0.025 mg/L and cannot, within the uncertainty limits, be distinguished from the total iron concentration. So much so that we can safely conclude that, in the iron citrate complex, iron is practically completely present as iron (II) and that the above spectrophotometric assays of iron support the composition $FeC_6H_6O_7 \cdot H_2O$ for the analysed iron citrate complex (since both the total iron concentration and the iron (II) concentration are, within their uncertainty, identical with the expected iron concentration).

The above solution of the iron complex was also tested for iron (III) using the thiocyanate method. In fact, thiocyanate is generally used for the detection and assay of iron (III) because of the formation of red iron (III) thiocyanate complexes, which absorb strongly at 480 nm. However, in agreement with the above results, no measurable absorption could be detected when samples of the solution of the iron citrate complex were treated, under controlled conditions, with a deaerated 15% *w/w* potassium thiocyanate solution.

Finally, the collected data show the exclusive presence of iron (II) ion in the synthesized iron citrate complex and this represents an advantage over other formulations of supplements that contain iron (III) ion. In fact, iron (II) is better assimilated (high bioavailability) compared to iron (III) [28].

For its use as an iron supplement, it is necessary to evaluate the stability at different pH of iron citrate complexes. For example, inside the human stomach the synthetic product is exposed at a very low pH and an evaluation of the stability at different pH allows one to understand whether the supplement could be taken as such or if it would be preferable to use a protective film for the stomach that would allow the dissolution and assimilation of iron in the intestine.

In this respect, spectrophotometric assays to determine the oxidation state of iron reported above were conducted for iron citrate solutions of different pH (from 2 to 12), confirming the stability of iron ion in the oxidation state +2 in the iron citrate complex. Furthermore, there was no precipitation of iron as ferrous hydroxide or ferric hydroxide in the evaluated pH range.

Two samples of iron citrate powder were subjected to X-ray analysis, obtaining a series of spectra that showed the presence of a crystalline-type material. Figure 2 shows the spectra, which confirm that the synthesis procedure allowed us to obtain a crystalline compound. Finally, the sample obtained in crystalline form was subjected to a re-crystallization test to try to obtain single crystals. However, until now, it has not been possible to obtain a well-defined single crystal.

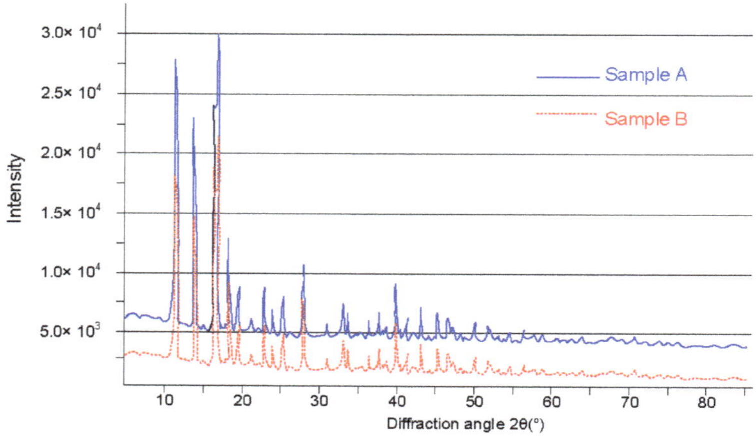

Figure 2. Comparison between X-ray diffraction spectra of two samples of iron citrate powder. Samples A and B are two independent replicates.

In order to verify the hypothetical coordination by the metal with the organic citrate type sites, an infrared spectroscopy (IR) analysis of a solid sample of citric acid and sodium citrate was carried out, stratifying the samples on the cell using ethanol. The ATR spectra of the samples of iron citrate were compared with those of citric acid and sodium citrate. Particularly interesting was the comparison of the spectra obtained for two samples of iron citrate powder, in particular from the peaks relative to the zone of the signals related to the COO^- showing a typical band around 1600/1700 cm^{-1}. The presence of metal-related peaks, which were found only for the samples of iron citrate (Figure 3), allowed us to establish the right coordination linking by the citrate with the iron.

Birsa Čelič et al. [24] previously reported the synthesis of a neutral carboxylated ferrous citrate compound that is similar to the iron (II) citrate complex that is presented in the current study. However, the carboxylated ferrous citrate polymer was hydrothermally synthesized by reacting ferrous chloride with citric acid in a basic solution. Consequently, ionized citric acid in solution is carboxylated by irregular bonds to iron (II), in which iron (II) is octahedrally complexed by different oxygen atoms. Complexation of $[Fe(H_2cit)(H_2O)]n$ times, and the crosslinking between iron (II) citric acid, resulted in the formation of an iron polymer compound. As confirmed by single-crystal X-ray diffraction, the structure of the compound can be described as a pseudo-three-dimensional framework composed of infinite chains of coordinated polyhedral iron. Indeed, the iron atoms are hexacoordinated by two partially deprotonated citrates and a water molecule in a distorted octahedral fashion. Furthermore, the reaction was carried out in an autoclave at a temperature of 150 °C (423 K) for one day and the polymer maintains its magnetic proprieties and stability at temperatures up to 275 °C (548 K). In contrast, an iron (II) citrate complex was formed by the reaction of iron filings and citric acid in 2 h in the current study. The diffraction spectra obtained from the two compounds cannot be superimposed, especially in the region between 40 and 45° (diffraction angle 2θ), thus indicating that the two compounds are different. The ferrous citrate compound reported in the literature showed the coordination of iron (II) in a complex octahedral structure, possibly as a long chain of polymers rather than a 1:1 coordination of iron ions and dibasic citrate in our compound, which was formed by the dehydrogenation of citrate with consequent formation of a neutral complex (see the reaction (1)). Finally, and most importantly, the compound that was reported earlier focused on the preparation of iron citrate metal complex with magnetic properties that could be potential candidates for biomedical applications rather than the synthesis of an iron compound that is to be used as a food supplement or an additive for the management of iron deficiency.

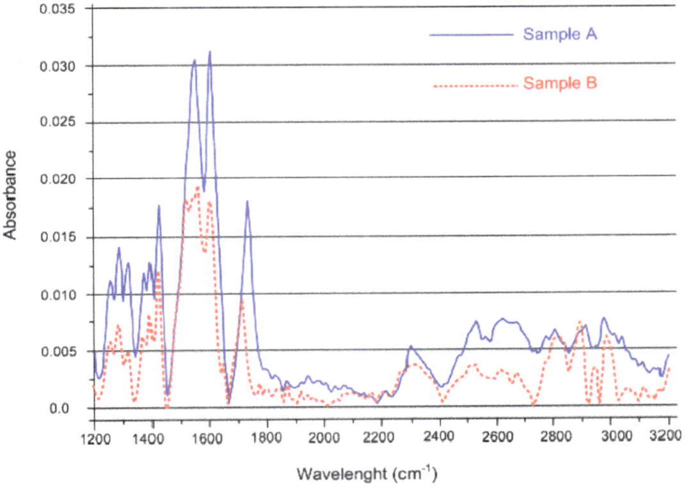

Figure 3. Comparison between IR spectra of two samples of iron citrate powder. Samples A and B are two independent replicates.

The good results obtained from the chemical characterization of the iron citrate complex led us to consider its potential application as an iron supplement. In fact, the enrichment of foods with iron citrate would allow individuals with iron deficiency to not change their eating habits and at the same time avoid taking pharmaceutical formulations to supplement this element in the body.

4. Conclusions

The current study shows a rapid and easy method to synthesise iron (II) citrate complex from iron filings and citric acid. The complex was subjected to a battery of analytical techniques to demonstrate that the synthesised crystalline solid corresponds to the formula $FeC_6H_6O_7 \cdot H_2O$. The relevance of the synthesised complex, exclusively containing iron in the +2 oxidation state, is associated with its potential use as an alternative to commercially available iron supplements and additives.

Further studies, some already in progress, will be conducted on the synthesised complex in order to obtain data with undoubted value for the realization of a food supplement. In this respect, further research is needed to validate the real effectiveness and mechanism of action of this complex. In vitro and in vivo studies will be performed, first on cells and animal models, in order to evaluate its safety and toxicity. Finally, the assimilation within the human body will be evaluated in human volunteers.

Furthermore, enrichment tests will be performed on various food matrices to evaluate the macroscopic effects of the iron (II) citrate complex on the enriched foods.

Author Contributions: D.N. and M.G. conceived and organized the manuscript; D.N., M.M.S. and M.G. wrote the text; M.L., M.C., C.L. and S.L. performed the analysis; D.N., M.M.S., A.A., F.S. and M.G. edited and reviewed the manuscript.

Funding: Financial support was provided by the Ministry of Education, University and Research (MIUR, Italy) through Finanziamento delle Attività Base della Ricerca (FFABR) 2017 assigned to Prof.ssa Monica Gallo.

Acknowledgments: We acknowledge Marco Trifuoggi, Antonio Carella, Carla Manfredi, Rocco Di Girolamo and dott. Antonio Polcaro (Polcaro Fitopreparazioni).

Conflicts of Interest: The authors declare no conflict of interest.

References

1. Muckenthaler, M.U.; Rivella, S.; Hentze, M.W.; Galy, B. A red carpet for iron metabolism. *Cell* **2017**, *168*, 344–361. [CrossRef] [PubMed]
2. De-Regil, L.M.; Jefferds, M.E.D.; Sylvetsky, A.C.; Dowswell, T. Intermittent iron supplementation for improving nutrition and development in children under 12 years of age. *Cochrane Database Syst. Rev.* **2011**, *12*, CD009085. [CrossRef] [PubMed]
3. World Health Organization. *Guideline: Daily Iron and Folic Acid Supplementation in Pregnant Women*; World Health Organization, 2012. Available online: http://apps.who.int/iris/bitstream/10665/77770/1/9789241501996_eng.pdf (accessed on 8 June 2012).
4. Peña-Rosas, J.P.; De-Regil, L.M.; Dowswell, T.; Viteri, F.E. Daily oral iron supplementation during pregnancy. *Cochrane Database Syst. Rev.* **2012**, *12*, CD004736. [PubMed]
5. Cantor, A.G.; Bougatsos, C.; Dana, T.; Blazina, I.; McDonagh, M. Routine iron supplementation and screening for iron deficiency anemia in pregnancy: A systematic review for the US Preventive Services Task Force. *Ann. Intern. Med.* **2015**, *162*, 566–576. [CrossRef] [PubMed]
6. Camaschella, C. Iron-deficiency anemia. *N. Engl. J. Med.* **2015**, *373*, 485–486. [CrossRef] [PubMed]
7. Bogdan, A.R.; Miyazawa, M.; Hashimoto, K.; Tsuji, Y. Regulators of iron homeostasis: New players in metabolism, cell death, and disease. *Trends Biochem. Sci.* **2016**, *41*, 274–286. [CrossRef] [PubMed]
8. Dixon, S.J.; Stockwell, B.R. The role of iron and reactive oxygen species in cell death. *Nat. Chem. Biol.* **2014**, *10*, 9. [CrossRef] [PubMed]
9. Julián-Almárcegui, C.; Heath, A.L.M.; Harvey, L.; Sarria, B.; Huybrechts, I. Intake and dietary sources of heme and non-heme iron in children and adolescents: Recommendations for preventing iron deficiency. In *Handbook of Nutrition and Diet in Leukemia and Blood Disease Therapy*; Wageningen Academic Publishers: Gelderaland, The Netherlands, 2016; pp. 371–398.
10. Pereira, P.M.D.C.C.; Vicente, A.F.D.R.B. Meat nutritional composition and nutritive role in the human diet. *Meat Sci.* **2013**, *93*, 586–592. [CrossRef] [PubMed]
11. Conrad, M.E.; Umbreit, J.N. Pathways of iron absorption. *Blood Cells Mol. Dis.* **2002**, *29*, 336–355. [CrossRef] [PubMed]
12. Weinborn, V.; Valenzuela, C.; Olivares, M.; Arredondo, M.; Weill, R.; Pizarro, F. Prebiotics increase heme iron bioavailability and do not affect non-heme iron bioavailability in humans. *Food Funct.* **2017**, *8*, 1994–1999. [CrossRef] [PubMed]
13. Abbaspour, N.; Hurrell, R.; Kelishadi, R. Review on iron and its importance for human health. *J. Res. Med. Sci.* **2014**, *19*, 164. [PubMed]
14. Crichton, R. *Iron Metabolism: From Molecular Mechanisms to Clinical Consequences*; John Wiley & Sons: Chichester, UK, 2016.
15. Hercberg, S.; Preziosi, P.; Galan, P. Iron deficiency in Europe. *Public Health Nutr.* **2001**, *4*, 537–545. [CrossRef] [PubMed]
16. Lopez, A.; Cacoub, P.; Macdougall, I.C.; Peyrin-Biroulet, L. Iron deficiency anaemia. *Lancet* **2016**, *387*, 907–916. [CrossRef]
17. Wu, A.C.; Lesperance, L.; Bernstein, H. Screening for iron deficiency. *Policy Statement* **2016**, *23*, 171–177. [CrossRef]

18. Dickinson, A.; MacKay, D. Health habits and other characteristics of dietary supplement users: A review. *Nutr. J.* **2014**, *13*, 14. [CrossRef] [PubMed]
19. Langella, C.; Naviglio, D.; Marino, M.; Gallo, M. Study of the effects of a diet supplemented with active components on lipid and glycemic profiles. *Nutrition* **2015**, *31*, 180–186. [CrossRef] [PubMed]
20. Rautiainen, S.; Manson, J.E.; Lichtenstein, A.H.; Sesso, H.D. Dietary supplements and disease prevention—A global overview. *Nat. Rev. Endocrinol.* **2016**, *12*, 407. [CrossRef] [PubMed]
21. Osredkar, J.; Sustar, N. Copper and zinc, biological role and significance of copper/zinc imbalance. *J. Clin. Toxicol.* **2011**, *3*. [CrossRef]
22. Martınez-Navarrete, N.; Camacho, M.M.; Martınez-Lahuerta, J.; Martınez-Monzó, J.; Fito, P. Iron deficiency and iron fortified foods—A review. *Food Res. Int.* **2002**, *35*, 225–231. [CrossRef]
23. United States Patent Office. Method of Making Ferrous Citrate. Patent 3,091,626, 28 May 1963.
24. Birsa Čelič, T.; Jagličič, Z.; Lazar, K.; Zabukovec Logar, N. Structure and magnetic properties of a new iron (II) citrate coordination polymer. *Acta Crystallogr. B Struct. Sci. Cryst. Eng. Mater.* **2013**, *69*, 490–495. [CrossRef] [PubMed]
25. Fortune, W.B.; Mellon, M.G. Determination of iron with o-phenanthroline: A spectrophotometric study. *Ind. Eng. Chem. Anal. Ed.* **1938**, *10*, 60–64. [CrossRef]
26. Lister, M.W.; Rivington, D.E. Some measurements on the iron (111)–thiocyanate system in aqueous solution. *Can. J. Chem.* **1955**, *33*, 1572–1590. [CrossRef]
27. Miller, J.; Miller, J.C. *Statistics and Chemometrics for Analytical Chemistry*, 6th ed.; Pearson Educational Limited: Horlow, UK, 2010.
28. Wessling-Resnick, M. Iron: Basic nutritional aspects. In *Molecular, Genetic, and Nutritional Aspects of Major and Trace Minerals*; Academic Press: London, UK, 2016; pp. 161–173.

nutrients

MDPI

Article

Iron Transport from Ferrous Bisglycinate and Ferrous Sulfate in DMT1-Knockout Human Intestinal Caco-2 Cells

Xiaonan Yu, Lingjun Chen, Haoxuan Ding, Yang Zhao and Jie Feng *

Key Laboratory of Animal Nutrition & Feed Science, Zhejiang Province, College of Animal Sciences, Zhejiang University, Hangzhou 310058, China; 21617026@zju.edu.cn (X.Y.); 21717018@zju.edu.cn (L.C.); 11717014@zju.edu.cn (H.D.); 21817006@zju.edu.cn (Y.Z.)

* Correspondence: fengj@zju.edu.cn; Tel.: +86-571-8898-2121

Received: 2 January 2019; Accepted: 21 February 2019; Published: 26 February 2019

Abstract: This experiment was conducted to investigate the transport characteristics of iron from ferrous bisglycinate (Fe-Gly) in intestinal cells. The divalent metal transporter 1 (DMT1)-knockout Caco-2 cell line was developed by Crispr-Cas9, and then the cells were treated with ferrous sulfate ($FeSO_4$) or Fe-Gly to observe the labile iron pool and determine their iron transport. The results showed that the intracellular labile iron increased significantly with Fe-Gly or $FeSO_4$ treatment, and this phenomenon was evident over a wide range of time and iron concentrations in the wild-type cells, whereas in the knockout cells it increased only after processing with high concentrations of iron for a long time ($p < 0.05$). DMT1-knockout suppressed the synthesis of ferritin and inhibited the response of iron regulatory protein 1 (IRP-1) and IRP-2 to these two iron sources. The expression of peptide transporter 1 (PepT1) was not altered by knockout or iron treatment. Interestingly, the expression of zinc-regulated transporter (ZRT) and iron-regulated transporter (IRT)-like protein 14 (Zip14) was elevated significantly by knockout and iron treatment in wild-type cells ($p < 0.05$). These results indicated that iron from Fe-Gly was probably mainly transported into enterocytes via DMT1 like $FeSO_4$; Zip14 may play a certain role in the intestinal iron transport.

Keywords: DMT1; knockout; ferrous bisglycinate; ferrous sulfate; transport; intestinal

1. Introduction

Iron is one of the essential trace elements and a cofactor for various enzymes. It is involved in many important physiological processes, such as oxygen transport, electron transport, tricarboxylic acid cycle, and hemoglobin and myoglobin production. Iron deficiency can induce anemia and affects cell metabolism [1–3]. Taking into account the vital functions of iron, exogenous iron is often supplemented as food fortification and feed additive to human and animals to prevent iron deficiency [4–6]. However, due to the low bioavailability of inorganic iron, it often causes mineral resources waste and environmental pollution, and its abuse also affects human health and animal production [7–9].

The use of more efficient sources of iron is a strategy to alleviate the problems caused by excess inorganic iron. Recent studies have shown that amino acid chelated iron sources such as Fe-Gly have better iron bioavailability than inorganic iron sources such as $FeSO_4$ [10–13]. The high bioavailability of amino acid chelated iron may be related to its efficient absorption and transport mechanism. However, at present, its intestinal absorption and transport mechanisms are still unclear, since there were few studies on this aspect. Some researchers thought that amino acid chelated iron might be transported across the brush border membrane (BBM) as a whole by the peptide transporter 1 (PepT1), similar to the intestinal absorption of small peptides. Their studies have shown that the intestinal uptake of amino acid chelated iron was in greater amounts and more rapid than equivalent quantities of iron salts. Furthermore, compared with $FeSO_4$, Fe-Gly has been shown to significantly increase PepT1 mRNA level and protein expression in the small intestinal epithelium cells (IPEC-1) of pigs [14,15]. Others thought that amino acid chelated iron was absorbed by the enterocytes in the same manner as inorganic iron. Researchers have demonstrated that iron from the Fe-Gly competed with $FeSO_4$ for the nonheme-iron absorption pathway [16]. Fe-Gly and $FeSO_4$ had similar absorption kinetics characteristics and their intestinal absorptions were significantly inhibited by divalent metal ions [17,18]. In our previous study, the duodenum of Sprague–Dawley (SD) rats given Fe-Gly or $FeSO_4$ were collected for transcriptome sequencing. The results showed that there was no significant difference in the expression of any iron transporter in the duodenum of rats administrated with Fe-Gly or $FeSO_4$, suggesting that the amino acid chelated iron may be absorbed by the intestine in the same manner as inorganic iron, which was mainly transported into the intestinal cells via the divalent metal ion transporter DMT1 [19].

DMT1 is a key mammalian iron transporter in the enterocytes, and it is selective for the ferrous iron [20,21]. Inorganic dietary iron exists in ferric iron form predominantly; it must first be reduced by duodenal cytochrome B (DcytB) and then enters the enterocytes via DMT1 located in the apical membrane of enterocytes [22–24].

To study the iron transport from Fe-Gly in the intestinal cells, in the present study, we developed a DMT1-knockout Caco-2 cell line by using Crispr-Cas9 and treated the cells with Fe-Gly and $FeSO_4$ to measure the changes of labile iron pool, ferritin content, and expression of iron regulators and transporters.

2. Materials and Methods

2.1. Cell Culture

Human Caco-2 cells (HTB-37) were obtained from American Type Culture Collection. Cells were cultured in 25-cm^2 flasks in Dulbecco's Modified Eagle Medium (DMEM) (Gibco, NYC, USA) with 10% fetal bovine serum (Gibco, NYC, USA), 100 U/mL penicillin-streptomycin (Gibco, NYC, USA) and 1% non-essential amino acids (Gibco, NYC, USA) and were incubated at 37 °C, 5% CO_2. Monolayers were grown by seeding Caco-2 cells (passage 20–40) at a density of 1×10^4 cells/cm^2 in 6-well plates or 96-well plates for about 14 days, with media changed every 2 days.

2.2. Knockout of DMT1 in Caco-2 Cells by Using Crispr Cas9

To generate CRISPR knockout cells, Caco-2 cells were transfected with plasmids expressing Cas9, Puro, Amp, and sgRNA for DMT1 and grown for 72 h. The sgRNAs used included 5′-AGAGAGGGATTACTATAGGCAGG-3′, 5′-CATGGGGAGTCTGCCAGTCTTGG-3′, and 5′-GAA GATCTCCATTCCTGAGGAGG-3′, which were designed based on the first exon of the public coding sequence (CDS) region located in all transcripts of DMT1. After that, the cells were screened by puromycin and separated as single cells into 96-well plates. After 2 weeks, the colonies were transferred to 48-well plates for expansion. After the cells were covered with 48-well plates, a portion (10^2–10^4) was removed and the Genloci TNA extraction kit (Genloci Biotechnologies Inc., Nanjing, China) was used to extract the cell genome. The following reaction system was prepared in a sterile Polymerase Chain Reaction (PCR) tube to amplify wild-type and mutant-type sufficiently hybridized DNA products. A pair of highly specific primers were designed near the knockout target site and the amplified product was approximately 316 bp in length. Primer sequences were as follows: 5′-GTATACTAAGGATGAATTGT-3′, 3′-AACCTGAGGCTGCTGAACTT-5′. The procedure used in the polymerase chain reaction (PCR) program was an initial denaturation (1.5 min at 95 °C), a three-step amplification program (10 s at 95 °C, 10 s at 62 °C, and 20 s at 72 °C) that was repeated 40 times, a complete extension (5 min at 95 °C) and a final denaturation (3 min at 72 °C). The amplified products were initially screened by the Cruiser™

Enzyme (Genloci Biotechnologies Inc., Nanjing, China) to obtain positive clones, and then the clones were further verified by sequencing and western blot analysis.

2.3. Measurement of Cell Viability

Wild-type and DMT1 knockout Caco-2 cells were seeded at a density of 1×10^4 cells/cm^2 in 96-well plates. After 14 days, the medium was discarded and the cells were washed twice with PBS and then incubated with DMEM containing 100 μM deferoxamine (DFO) (Sigma, St. Louis, MO, USA). Twenty-four hours later, the medium was discarded and the cells were rinsed twice with PBS before being treated with DMEM containing different concentrations of Fe-Gly and $FeSO_4$ (0, 25, 50, 100, 200 μM). Two hours later, the medium was discarded and the cell viability was assessed by Cell Counting Kit-8 (MedChem Express, Princeton, NJ, USA).

2.4. Measurement of Labile Iron with Phen Green SK in Wild-Type and DMT1 Knockout Caco-2 Cells

Wild-type and DMT1 knockout Caco-2 cells were seeded at a density of 1×10^4 cells/cm^2 in 96-well plates. After 14 days, the medium was discarded and the cells were washed twice with PBS and then incubated with DMEM containing 100 μM DFO. Twenty-four hours later, the medium was discarded, and the cells were rinsed twice with PBS before being treated with DMEM containing different concentrations (0, 25, 50, 100, 200 μM) of Fe-Gly (Dibo Biotechnology Co., Ltd., Shanghai, China) and $FeSO_4$ (Sigma, St. Louis, MO, USA). The cells were incubated for 0, 0.5, 1, 2 h and then rinsed twice with PBS and stained with PBS containing 10 μM Phen Green SK (Thermo Fisher P14313) for 1 h at 37 °C. After staining, cells were rinsed with PBS twice and labile iron was measured by using a fluorescence microplate reader (SpectraMax M5, Molecular Devices, Sunnyvale, CA, USA) at an excitation wavelength of 490 nm and an emission wavelength of 520 nm.

2.5. Live Fluorescence Imaging of Labile Iron in Wild-Type and DMT1 Knockout Caco-2 Cells

To visualize labile iron, microscopy imaging of the fluorescent dye Phen Green SK (Invitrogen, Eugene, OR, USA) was performed. Wild-type and DMT1 knockout Caco-2 cells were seeded at a density of 1×10^4 cells/cm^2 in 6-well plates. After 14 days, the medium was discarded and the cells were washed twice with PBS and then incubated with DMEM containing 100 μM DFO. Twenty-four hours later, the medium was discarded and the cells were rinsed twice with PBS before being treated with DMEM containing different concentrations of Fe-Gly and $FeSO_4$ (0, 25, 50, 100, 200 μM). The cells were incubated for an additional 2 h, then rinsed twice with PBS and stained with PBS containing 10 μM Phen Green SK for 1 h at 37 °C. Next, the stained cells were rinsed with PBS and observed with a fluorescence microscope (IX7, Olympus Corporation, Tokyo, Japan).

2.6. Western Blot Analysis

After wild-type and DMT1 knockout Caco-2 cells in the 6-well plates were pretreated with DFO as described above, the cells were rinsed twice with PBS and then treated with DMEM containing 0 or 25 μM Fe-Gly or $FeSO_4$ for 2 h. Then cells were lysed with ice-cold radio immunoprecipitation assay (RIPA) buffer (Beyotime Biotechnology, Shanghai, China) containing protease inhibitors and phenylmethanesulfonyl fluoride (PMSF, Beyotime Biotechnology, Shanghai, China) for about 1 min on ice. The lysates were transferred to a 1.5-mL centrifuge tube and were centrifuged at 11,000 × *g* for 10 min at 4 °C, then the supernatants were collected to determine the total protein concentrations using a BCA Protein Assay kit (Keygen biotech. Co. Ltd., Nanjing, China). Next, 5X dual color protein loading buffer (FD bioscience, Hangzhou, China) was added to the supernatant and then the samples were boiled for protein extraction. The extracted proteins (20–40 μg) were separated by electrophoresis on a 10% SDS-PAGE gel and transferred onto an activated polyvinylidene fluoride (PVDF) membrane (GE Healthcare Life science, Germany). Subsequently, the membrane was blocked in 5% non-fat milk at room temperature for 1 or 2 h and then incubated overnight at 4 °C with the following primary antibodies and dilution rates: DMT1, 1:500 (Santa Cruz Biotechnology, code sc-166884, Santa Cruz, CA,

USA); Ferritin, 1:1000 (Abcam, code ab75973, Cambridge, UK); iron regulatory protein 1 (IRP-1), 1:1000 (Abcam, code ab126595, Cambridge, UK); IRP-2, 1:400 (Proteintech Group, code23829-1-AP, Chicago, IL, USA); hypoxia-induced factor-2α (HIF-2α), 1:1000 (Abcam, code ab207607, Cambridge, UK); PepT1, 1:200 (Abcam, code ab123314, Cambridge, UK); ferroportin 1 (FPN1), 1:2000 (Proteintech Group, code 26601-1-AP, Chicago, IL, USA); iron-regulated transporter (IRT)-like protein 14 (Zip14), 1:500 (Abcam, code ab106568, Cambridge, UK); and β-Actin, 1:2000 (Bioker biotechnology, code BK-7018, Hangzhou, China). Then the membrane was rinsed for 10 min three times thoroughly with TBST before incubation with secondary antibody consisting of goat anti-rabbit (1:20,000, Bioler biotechnology, code BK-R050) and goat anti-mouse (1:20,000, Bioker biotechnology, code BK-M050, Hangzhou, China) at room temperature for about 2 h. After that, the membrane was thoroughly rinsed with TBST for 10 min three times. The signals were detected after the addition of ECL Star Chemiluminescence solution according to the manufacturer's instructions (Beyotime Biotechnology, Shanghai, China).

2.7. Statistical Analysis

All data are presented as the means or weighted means ± SEM of a minimum of three biological replicates unless otherwise noted. Means between groups were compared by one-way analysis of variance and post-hoc Tukey test or non-parameter Kruskal-Wallis test (SPSS software, version 21, SPSS Inc., Chicago, IL, USA) where appropriate. For this study, $p < 0.05$ was considered significant.

3. Results

3.1. Knockout of DMT1 in Caco-2 Cells by Using Crispr Cas9

To verify the targeted disruption of DMT1 in Caco-2 cells by the Crispr-Cas9 system, we analyzed genomic DNA isolated from transfected cells using Cruiser^{TM} Enzyme assay. A 316-base pair (bp) sequence flanking the target site treated by sgRNA-encoded plasmids was amplified by PCR. As expected, the lengths of the PCR products were obviously shorter in mutant cell clones (Figure 1A). Sequencing analysis of the PCR products of these clones revealed that the mutant cells showed 85-bp deletions (5'-TATAGTAATCCCTCTCTTTCACAGTCCCCTGGGGACTCAGAGGAGTACTTCGCCAC TTACTTTAATGAGAAGATCTCCATTCCTG-3') on the exon from the DMT1 gene (Figure 1B–D). Therefore, the mutant was a positive knockout cell line on the genome. We further verified the DMT1 mutation on protein expression level. Western blot results (Figure 1E) showed that there was almost no protein expression of DMT1 in #30–125, which confirmed that the DMT1 knockout Caco-2 cell line was successfully developed.

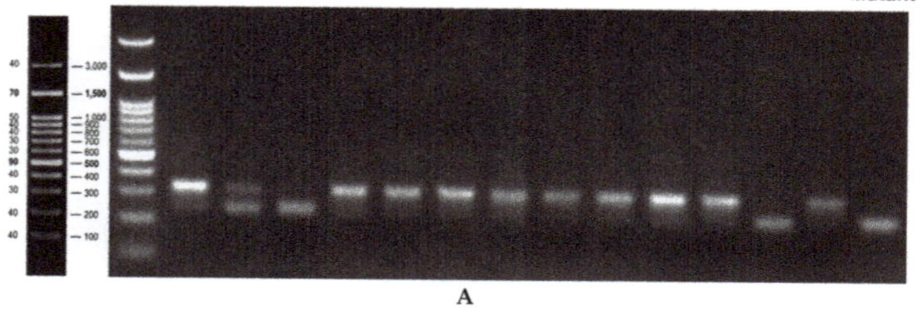

A

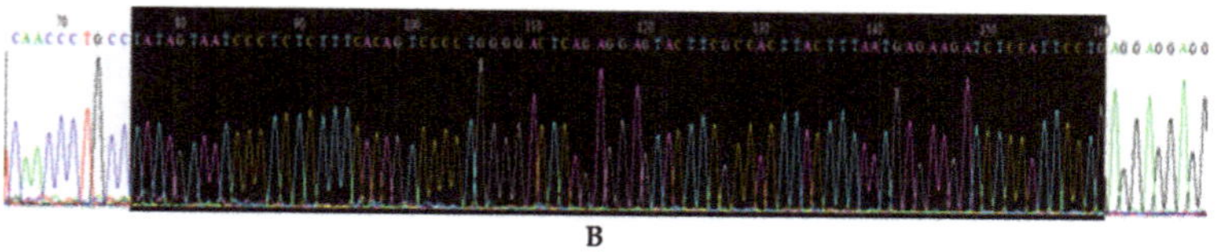

B

Figure 1. *Cont.*

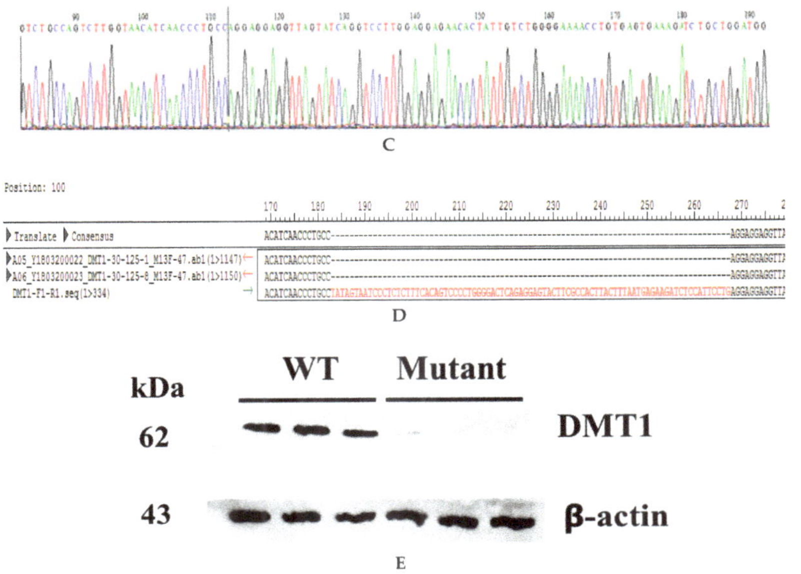

Figure 1. Validation of DMT1-knockout Caco-2 cell line. (**A**) The electrophoresis results of the target fragments of DMT1 in the transfected cells; (**B**) Partial sequencing results of the target fragment on DMT1 of wild-type Caco-2 cells; (**C**) Partial sequencing results of the target fragment on DMT1 of the mutant cells; (**D**) Sequence comparison of the target fragment of DMT1 in the mutant and wild-type Caco-2 cells; (**E**) Western blot results of DMT1 in wild-type Caco-2 cells and the mutant cells.

3.2. Cell Viability after 2 h of Iron Treatment

As shown in Figure 2, the treatment of $FeSO_4$ or Fe-Gly at concentration from 25 μM to 200 μM for 2 h did not affect the viability of wild-type and DMT1 knockout Caco-2 cells.

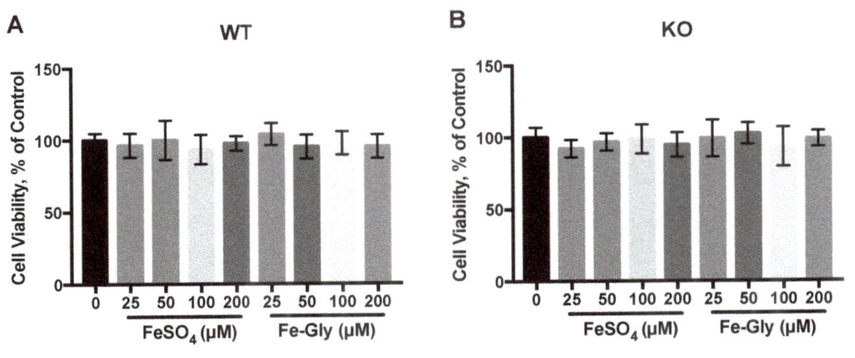

Figure 2. WT: wild-type Caco-2 cell; KO: DMT1-knockout Caco-2 cell. Cell viability of (**A**) wild-type and (**B**) DMT1-knockout Caco-2 cells after 2 h of iron treatment.

3.3. Changes of Labile Iron after Treatment with Different Iron Sources

The relative labile iron level was determined by using the turn-off probe Phen Green SK, which was quenched upon intracellular iron binding. After treatment with $FeSO_4$ and Fe-Gly at different concentrations, the labile iron levels of wild-type Caco-2 cells and DMT1 knockout Caco-2 cells were recorded, as shown in Figure 3.

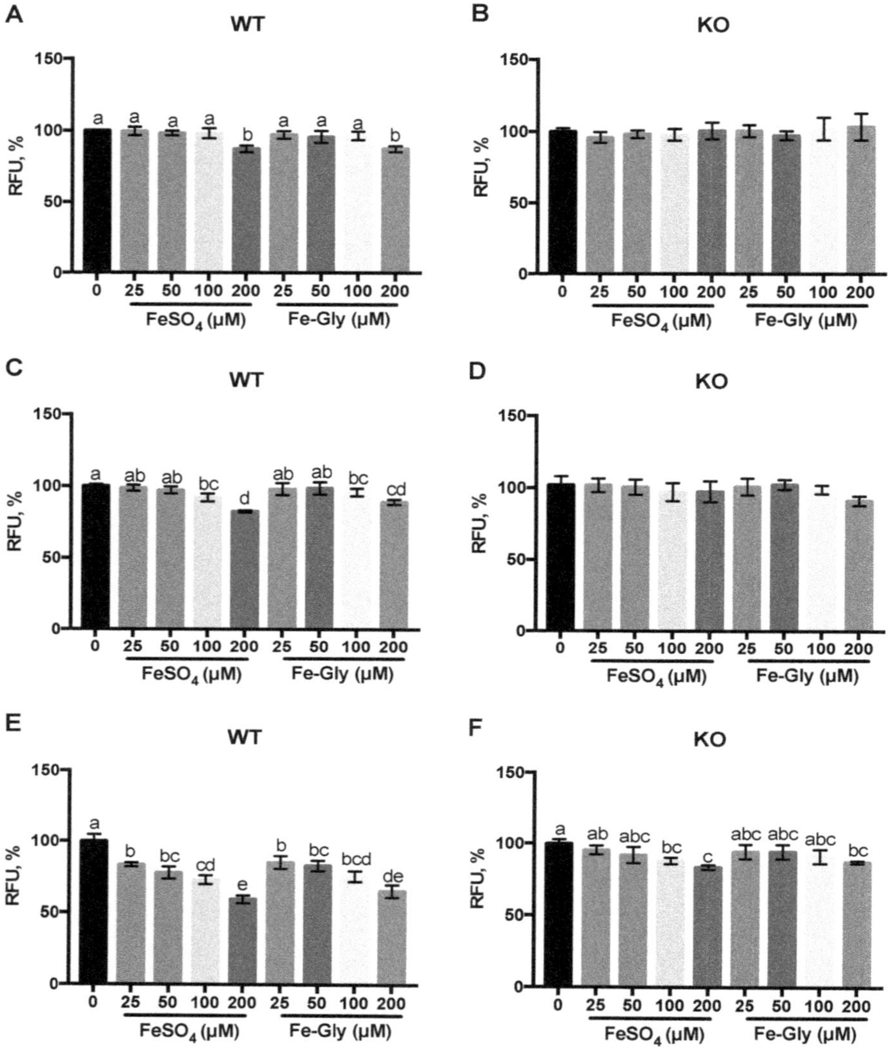

Figure 3. Changes of labile iron after treatment with different iron sources (0, 25, 50, 100, 200 μM) in wild-type cells and DMT1-knockout cells. Values not sharing a common letter differ significantly ($p < 0.05$). WT: wild-type Caco-2 cell; KO: DMT1-knockout Caco-2 cell; RFU: relative florescence unit. Changes of labile iron after 0.5 h (**A**,**B**), 1 h (**C**,**D**), 2 h (**E**,**F**) treatment with different iron sources in wild-type cells and DMT1-knockout cells.

After 30 min, 200 μM $FeSO_4$ or Fe-Gly lead to significant Phen Green SK quenching in the wild-type Caco-2 cells ($p = 0.000301$, $p = 0.000319$), which indicated the significant increase of labile iron, whereas the treatment of iron had no effect on it in the knockout cells (Figure 3A,B). An hour later, the labile iron level of wild-type cells began to increase when the concentration of $FeSO_4$ or Fe-Gly reached 100 μM ($p < 0.05$), while there was no change in it in the knockout cells (Figure 3C,D). Figure 3E,F show that there was a significant increase in the labile iron level in wild-type cells after treatment with 25, 50, 100, 200 μM $FeSO_4$ or Fe-Gly for two hours, while that in the knockout cells elevated after processing with 100 and 200 μM iron ($p < 0.05$).

3.4. Live Cell Fluorescence Imaging of Labile Iron with Phen Green SK

To visualize labile iron, cells on a 6-well plate were stained with Phen Green-SK and observed by a fluorescence microscopy. The cells were treated with gradient concentrations of iron for 2 h based on the results above. It can be seen from Figure 4A,B that low concentrations of Fe-Gly or $FeSO_4$ result in the quenching of Phen Green SK in the wild-type cells, which indicates the increase of labile iron, while that in the knockout cells began to change after processing with higher concentrations of iron. The results of the live cell fluorescence imaging of labile iron confirmed the results we obtained by the fluorescence microplate reader.

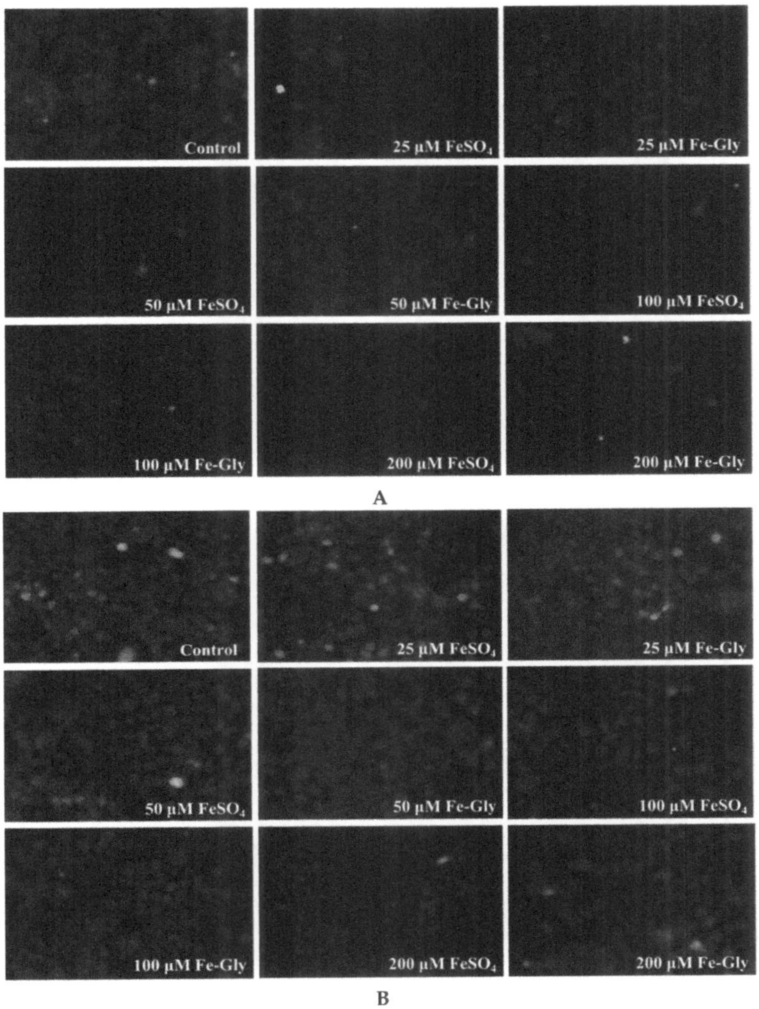

Figure 4. Live cell fluorescence imaging (200x) of labile iron in wild-type cells and DMT1-knockout cells after treated with different concentrations of $FeSO_4$ and Fe-Gly for two hours. (**A**) Fluorescence imaging of labile iron in wild-type cells; (**B**) Fluorescence imaging of labile iron in DMT1-knockout cells.

3.5. Expression of Iron Regulators and Transporters

The protein expression levels of the iron regulators and transporters were measured after cells were treated with 25 μM Fe-Gly or $FeSO_4$ for 2 h and normalized to β-actin. Obviously increased expression of HIF-2α, IRP-1, and IRP-2 and decreased content of ferritin was observed after DMT1-knockout in Caco-2 cells ($p < 0.05$) (Figure 5). Furthermore, treatment of these two iron sources increased ferritin content and decreased DMT1 expression levels significantly in the wild-type cells, while those regulators in the knockout cells did not respond to the Fe-Gly or $FeSO_4$ and there was almost no expression of DMT1 in the knockout cells (Figure 5). In the wild-type cells, the expression of IRP-1 was decreased significantly after treatment with $FeSO_4$ (Figure 5D). Figure 5E,F show that the expression of PepT1 and FPN1 were not altered by knockout or iron treatment. Interestingly, the expression of Zip14 was elevated significantly by knockout and iron treatment in wild-type cells ($p < 0.05$) and decreased slightly after iron treatment in knockout cells (Figure 5H).

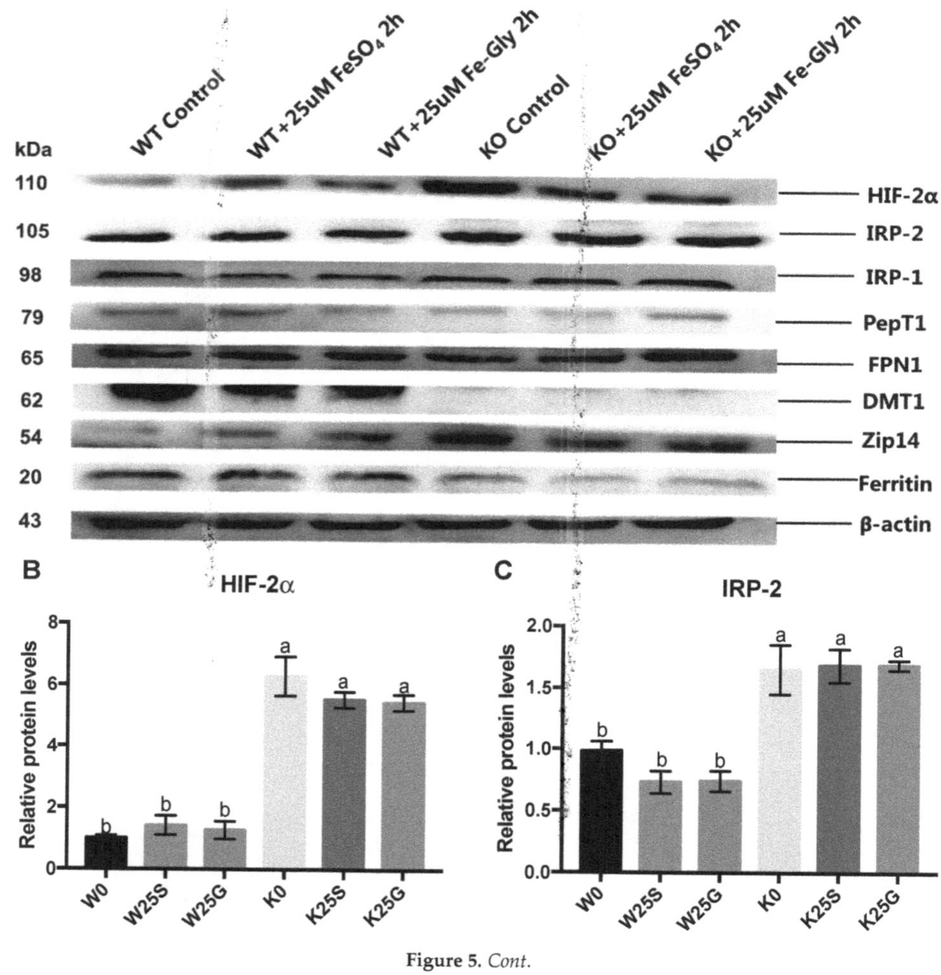

Figure 5. *Cont.*

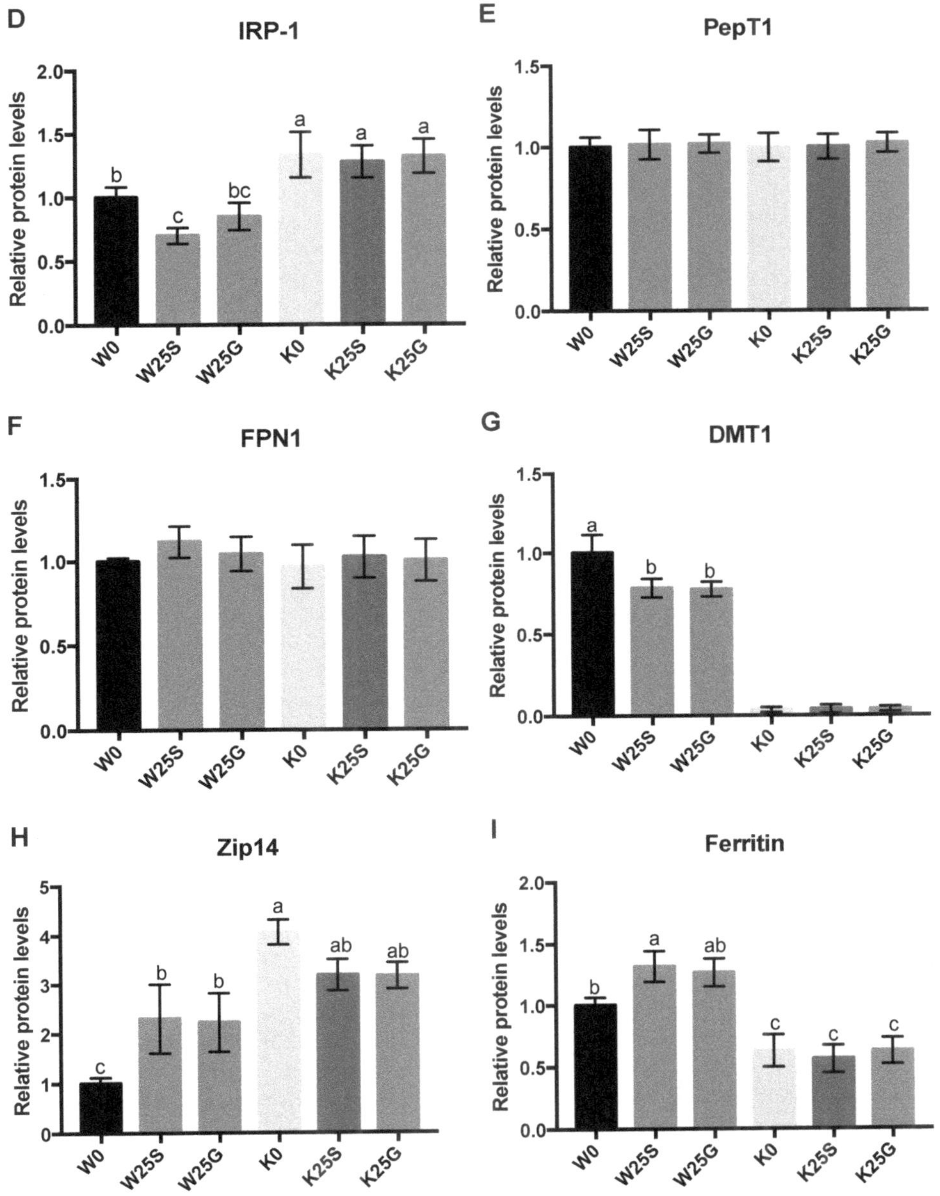

Figure 5. Expression of some iron regulators and transporters after treated with 25 μM of different iron sources for two hours. Values not sharing a common letter differ significantly ($p < 0.05$). W: wild-type cells treated with 0 μM iron; W25S, W25G: wild-type cells treated with 25 μM $FeSO_4$, 25 μM Fe-Gly; K0: DMT1-knockout cells treated with 0 μM iron; K25S, K25G: W0: wild-type cells treated with 0 μM iron; W25S, W25G: wild-type cells treated with 25 μM $FeSO_4$, 25 μM Fe-Gly; treated with 25 μM $FeSO_4$, 25 μM Fe-Gly. (**A**) Western blot bands of iron regulators and transporters. Statistical analysis of (**B**) hypoxia-induced factor-2α (HIF-2α), (**C**) iron regulatory protein 1 (IRP-1), (**D**) IRP-2, (**E**) peptide transporter 1 (PepT1), (**F**) ferroportin 1 (FPN1), (**G**) DMT1, (**H**) iron-regulated transporter (IRT)-like protein 14 (Zip14), (**I**) ferritin expression.

4. Discussion

The transport characteristics of Fe-Gly in the enterocytes have not been clarified yet. In this study, to further investigate whether iron from Fe-Gly was transported into enterocytes like inorganic iron via DMT1, we developed a DMT1-knockout Caco-2 cell line by using Crispr-Cas9 and treated the cells with Fe-Gly and $FeSO_4$ to observe the labile iron pool and determine the transport of the two forms of iron. DMT1 is the main transporter of inorganic iron in the intestine. After iron is transported into enterocytes through DMT1, it will first enter the cytosolic pool termed "labile iron pool", which is destined for metabolism, storage, or export [23,24]. The results we obtained by using the turn-off probe Phen Green SK showed that the intracellular labile iron increased significantly with Fe-Gly or $FeSO_4$ treatment, and this phenomenon was evident over a wide range of time and iron concentrations in the wild-type cells. However, in the knockout cells, the labile iron increased only after processing with high concentrations of iron for a long time ($p < 0.05$) (Figures 3 and 4). These results indicated that the increase of intracellular labile iron pool after treatment with Fe-Gly or $FeSO_4$ was inhibited by DMT1 knockout.

Iron from the labile iron pool that is not utilized or exported is stored in a non-toxic form within ferritin [1,2]. In this experiment, knockout of DMT1 decreased ferritin content ($p < 0.05$), and treatment with 25 μM Fe-Gly or $FeSO_4$ for 2 h increased its content in the wild-type cells ($p < 0.05$) but not in the DMT1-knockout cells (Figure 5I). The results indicated that the synthesis of ferritin was suppressed by DMT1 knockout. Since ferritin synthesis reflects intracellular iron bioavailability [2,6], the greater synthesis of ferritin in the wild-type cells suggests that the iron storage and utilization from Fe-Gly were affected by the absence of DMT1 as $FeSO_4$.

Cellular iron homeostasis is mainly orchestrated post-transcriptionally by iron regulatory proteins IRP-1 and IRP-2. The two orthologous RNA-binding proteins sense the labile iron pool, and the conversion of IRP-1 to aconitine and degradation of IRP-2 will be promoted upon iron replete [25,26]. In the current study, both IRP-1 and IRP-2 expression were increased due to the absence of DMT1 ($p < 0.05$), which reflected the lower intracellular iron content in the knockout cells than the wild-type cells. The results also showed that in wild-type cells, the protein expression level of IRP-1 decreased significantly after being treating with $FeSO_4$ and decreased slightly after being treated with Fe-Gly. The protein expression levels of IRP-2 decreased after treatment with both iron sources, but the difference was not significant. These results indicate the elevation of the intracellular iron levels in the cells. However, in the knockout cells, they were not altered by treatment with either of these two iron sources (Figure 5B,C). These results demonstrated that iron uptake from Fe-Gly and $FeSO_4$ were suppressed by DMT1 deficiency. IRPs interact with iron regulatory elements (IREs), which are present in the 5′ or 3′ untranslated regions (UTRs) of target mRNAs. IRP-binding to IREs responds to intracellular iron levels. During iron deficiency, IRPs bind to 5′UTR IRE of mRNAs encoding genes like HIF-2α and FPN1, thus inhibiting their translation. The IRPs also appear to stimulate the translation of genes like DMT1 when bound to the 3′ UTR IRE of their mRNAs [27–31]. In this study, the expression of FPN1 did not change significantly by knockout or iron treatment (Figure 5F). In the current study, the cells were cultured on an ordinary well plate, which may affect the basolateral transport of iron and thus may have affected the regulation of FPN1. In addition, some isoforms of FPN1 mRNA lack IRE and thus escape the IRP regulation [2]. HIF-2α did not respond to iron treatment either, but its expression was significantly increased by DMT1 knockout ($p < 0.05$) (Figure 5B). In addition to the regulation of the IRP system, HIF-2α is directly affected by iron content, and it will be degraded after iron-mediated proline hydroxylation. Under these two different regulatory mechanisms, there was a significant increase in HIF-2α in DMT1-knockout cells [32–34]. It was also observed in this experiment that after treatment with the two iron sources, the expression of DMT1 was significantly decreased in wild-type cells ($p < 0.05$), while there was almost no expression in knockout cells (Figure 5G). All of the above results indicated that the iron content in the knockout cells was lower than that of the wild-type cells, and the iron absorption of the two iron sources were both affected by the knockout of DMT1.

In addition, we also measured the expression of intestinal peptide transporter PepT1. A previous study reported that Fe-Gly significantly increases PepT1 mRNA level and protein expression in the small intestinal epithelium cells (IPEC-1) of pigs, and the reporter thought that Fe-Gly may be mediated into the intestinal cells via PepT1 [15]. However, in our study, the expression of PepT1 did not differ between wild-type and DMT1-knockout cells and between Fe-Gly and $FeSO_4$ treatment (Figure 5E). This result was consistent with our previous study, which showed that there was no difference in the expression of PepT1 in the duodenum of rats administrated with Fe-Gly or $FeSO_4$ [19].

Interestingly, when measuring the labile iron in DMT1 knockout cells, we found that the labile iron could elevate after treating with high concentrations of the two iron sources for a certain period of time. Since DMT1 was almost absent in the knockout cells and did not respond to iron treatment, we suspect that there may be other ways to mediate iron into the intestinal cells, and the transporter may be zinc-regulated transporter (ZRT) or iron-regulated transporter (IRT)-like protein 14 (Zip14). Zip14 was first proven to be a zinc transporter and recent studies have shown that it can mediate the absorption of non-transferrin bound iron (NTBI) in liver and other organs. NTBI appears the when the iron-binding capacity of plasma transferrin is exceeded under iron overload [35–40]. Recently, some studies have reported that it plays an important role in maintaining manganese homeostasis [41–43]. In the current study, we determined the protein expression of Zip14 and we found some interesting results. Knockout of DMT1 significantly increased the expression of Zip14 ($p < 0.05$). Furthermore, iron treatment significantly promoted Zip14 expression in wild-type cells ($p < 0.05$), while it slightly inhibited its expression in knockout cells (Figure 5H). Since DMT1 and Zip14 both can mediate the cellular uptake of metal ions, the high expression of Zip14 in knockout cells may be a complement to the lack of DMT1. Its elevation after iron treatment in wild-type cells may be due to the decreased expression of DMT1, which affected the absorption of other ions [38,39,44]. In knockout cells, its slight reduction may be in response to the iron treatment. These results supported that Zip14 may play a role in iron transport in intestinal cells. Several reports showed that the expression of Zip14 is relatively high in the intestine [38,45], but there have been few reports on its role in the intestine up until now. Here, the results of our experiments may provide some clues for further explorations and future studies.

5. Conclusions

The iron transport and storage from Fe-Gly in intestinal cells was affected by the knockout of DMT1 as well as $FeSO_4$, which indicated that Fe-Gly was probably mainly transported into the intestine cells via DMT1 like $FeSO_4$. Zip14 may play a certain role in intestinal iron transport.

Author Contributions: Conceptualization: X.Y. and J.F.; Methodology: X.Y. and J.F.; Validation: L.C., H.D., and Y.Z.; Formal analysis: X.Y. and L.C.; Investigation: X.Y.; Resources: J.F.; Data Curation: H.D.; Writing—Original Draft Preparation: X.Y.; Writing—Review and Editing: L.C. and J.F.; Visualization: X.Y.; Supervision: J.F.; Project Administration: J.F.; Funding Acquisition: J.F.

Funding: This research was funded by the National Natural Science Foundation of China (No.31772607 and 31472102) and Zhejiang Provincial Key R & D Project (2019C02051).

Acknowledgments: The authors would like to thank Shenglin Fang and Zhao Zhuo for their suggestions on the experimental method and would also like to thank Genloci Biotechnologies Inc. (Nanjing, China) for providing technical support for cell knockout.

Conflicts of Interest: The authors declare no conflict of interest.

References

1. Matthias, W.H.; Martina, U.M.; Bruno, G.; Clara, C. Two to Tango: Regulation of Mammalian Iron Metabolism. *Cell* **2010**, *142*, 24–38.
2. Martina, U.M.; Stefano, R.; Matthias, W.H.; Bruno, G. A Red Carpet for Iron Metabolism. *Cell* **2017**, *168*, 344–361.
3. Camaschella, C. Iron-deficiency anemia. *N. Engl. J. Med.* **2015**, *373*, 485–486. [CrossRef] [PubMed]

4. Li, Y.; Yang, W.; Dong, D.; Jiang, S.; Yang, Z.; Wang, Y. Effect of different sources and levels of iron in the diet of sows on iron status in neonatal pigs. *Anim. Nutr.* **2018**, *4*, 197–202. [CrossRef] [PubMed]
5. Creech, B.L.; Spears, J.W.; Flowers, W.L.; Hill, G.M.; Lloyd, K.E.; Armstrong, T.A.; Engle, T.E. Effect of dietary trace mineral concentration and source (inorganic vs. chelated) on performance, mineral status, and fecal mineral excretion in pigs from weaning through finishing. *J. Anim. Sci.* **2004**, *82*, 2140–2147. [CrossRef] [PubMed]
6. Antonio, P.; Christine, E.; Eugenia, V.; Paul, S.; Florentine, H.; Susan, F. Mechanisms of iron uptake from ferric phosphate nanoparticles in human intestinal Caco-2 cells. *Nutrients* **2017**, *9*, 359–372.
7. Feng, J.; Ma, W.Q.; Xu, Z.R.; He, J.X.; Wang, Y.Z.; Liu, J.X. The effect of iron glycine chelate on tissue mineral levels, fecal mineral concentration, and liver antioxidant enzyme activity in weanling pigs. *Anim. Feed Sci. Technol.* **2009**, *150*, 106–113. [CrossRef]
8. Kwiecien, M.; Samolinska, W.; Bujanowicz-Haras, B. Effects of iron–glycine chelate on growth, carcass characteristic, liver mineral concentrations and haematological and biochemical blood parameters in broilers. *J. Anim. Physiol. Anim. Nutr.* **2015**, *99*, 1184–1196. [CrossRef] [PubMed]
9. Chary, N.S.; Kamala, C.T.; Raj, D.S.S. Assessing risk of heavy metals fromconsuming food grown on sewage irrigated soils and food chain transfer. *Ecotox. Environ. Saf.* **2008**, *69*, 513–524. [CrossRef] [PubMed]
10. Feng, J.; Ma, W.Q.; Xu, Z.R.; Wang, Y.Z.; Liu, J.X. Effects of iron glycine chelate on growth, haematological and immunological characteristics in weaning pigs. *Anim. Feed Sci. Technol.* **2007**, *134*, 261–272. [CrossRef]
11. Ma, W.Q.; Sun, H.; Zhou, Y.; Wu, J.; Feng, J. Effects of iron glycine chelate on growth, tissue mineral concentrations, fecal mineral excretion, and liver antioxidant enzyme activities in broilers. *Biol. Trace Elem. Res.* **2012**, *149*, 204–211. [CrossRef] [PubMed]
12. Zhuo, Z.; Yu, X.; Li, S.; Fang, S.; Feng, J. Heme and non-heme iron on growth performances, blood parameters, tissue mineral concentration, and intestinal morphology of weanling pigs. *Biol. Trace Elem. Res.* **2018**. [CrossRef] [PubMed]
13. Yu, B.; Huang, W.; Chiou, P.W. Bioavailability of iron from amino acid complex in weanling pigs. *Anim. Feed Sci. Technol.* **2000**, *86*, 39–52. [CrossRef]
14. Ashmead, H.D. Comparative Intestinal Absorption and subsequent metabolism of metal amino acid chelates and inorganic metal salts. *Biol. Trace Elem. Res.* **1993**, *445*, 306–319.
15. Liao, Z.C.; Guan, W.T.; Chen, F.; Hou, D.X.; Wang, C.X.; Lv, Y.T.; Qiao, H.Z.; Chen, J.; Han, J.H. Ferrous bisglycinate increased iron transportation through DMT1 and PepT1 in pig intestinal epithelial cells compared with ferrous sulphate. *J. Anim. Feed Sci.* **2014**, *23*, 153–159. [CrossRef]
16. Fernando, P.; Manuel, O.; Eva, H.; Dora, I.M.; Miguel, A.; Angélica, L.; Virginia, G. Iron bis-glycine chelate competes for the nonheme-iron absorption pathway. *Am. J. Clin. Nutr.* **2002**, *76*, 577–581.
17. Yeung, C.K.; Glahn, R.P.; Miller, D.D. Inhibition of Iron Uptake from Iron Salts and Chelates by Divalent Metal Cations in Intestinal Epithelial Cells. *J. Agric. Food Chem.* **2005**, *53*, 132–136. [CrossRef] [PubMed]
18. Zhuo, Z.; Fang, S.; Yue, M.; Zhang, Y.; Feng, J. Kinetics absorption characteristics of ferrous glycinate in SD rats and its impact on the relevant transport protein. *Biol. Trace Elem. Res.* **2014**, *158*, 197–202. [CrossRef] [PubMed]
19. Zhuo, Z.; Fang, S.; Hu, Q.; Huang, D.; Feng, J. Digital gene expression profiling analysis of duodenum transcriptomes in SD rats administered ferrous sulfate or ferrous glycine chelate by gavage. *Sci. Rep.* **2016**, *6*, 37923. [CrossRef] [PubMed]
20. Fleming, M.D.; Trenor, C.C.; Su, M.A.; Foernzler, D.; Beier, D.R.; Dietrich, W.F.; Andrews, N.C. Microcytic anaemia mice have a mutation in Nramp2, a candidate iron transporter gene. *Nat. Genet.* **1997**, *16*, 383–386. [CrossRef] [PubMed]
21. Shawki, A.; Anthony, S.R.; Nose, Y.; Engevik, M.A.; Niespodzany, E.J.; Barrientos, T.; Öhrvik, H.; Worrell, R.T.; Thiele, D.J.; Mackenzie, B. Intestinal DMT1 is critical for iron absorption in the mouse but is not required for the absorption of copper or manganese. *Am. J. Physiol. Gastrointest. Liver Physiol.* **2015**, *309*, 635–647. [CrossRef] [PubMed]
22. Gunshin, H.; Mackenzie, B.; Berger, U.V.; Gunshin, Y.; Romero, M.F.; Boron, W.F.; Nussberger, S.; Gollan, J.L.; Hediger, M.A. Cloning and characterization of a mammalian proton-coupled metal-ion transporter. *Nature* **1997**, *388*, 482–488. [CrossRef] [PubMed]
23. Evstatiev, R.; Gasche, C. Iron sensing and signalling. *Gut* **2012**, *61*, 933–952. [CrossRef] [PubMed]

24. Bart, J.C.; Twan, L.; Stefano, R. Targeting iron metabolism in drug discovery and delivery. *Nat. Rev. Drug Discov.* **2017**, *16*, 400–423.
25. Salahudeen, A.A.; Thompson, J.W.; Ruiz, J.C.; Ma, H.W.; Kinch, L.N.; Li, Q.; Grishin, N.V.; Bruick, R.K. An E3 ligase possessing an iron- responsive hemerythrin domain is a regulator of iron homeostasis. *Science* **2009**, *326*, 722–726. [CrossRef] [PubMed]
26. Vashisht, A.A.; Zumbrennen, K.B.; Huang, X.; Powers, D.N.; Durazo, A.; Sun, D.; Bhaskaran, N.; Persson, A.; Uhlen, M.; Sangfelt, O.; et al. Control of iron homeostasis by an iron-regulated ubiquitin ligase. *Science* **2009**, *326*, 718–721. [CrossRef] [PubMed]
27. Muckenthaler, M.U.; Galy, B.; Hentze, M.W. Systemic iron homeostasis and the iron-responsive element/iron-regulatory protein (IRE/IRP) regulatory network. *Annu. Rev. Nutr.* **2008**, *28*, 197–213. [CrossRef] [PubMed]
28. Galy, B.; Ferring-Appel, D.; Kaden, S.; Gröne, H.J.; Hentze, M.W. Iron regulatory proteins are essential for intestinal function and control key iron absorption molecules in the duodenum. *Cell Metab.* **2008**, *7*, 79–85. [CrossRef] [PubMed]
29. Sanchez, M.; Galy, B.; Schwanhaeusser, B.; Blake, J.; Bähr-Ivacevic, T.; Benes, V.; Selbach, M.; Muckenthaler, M.U.; Hentze, M.W. Iron regulatory protein-1 and -2: Transcriptome-wide definition of binding mRNAs and shaping of the cellular proteome by iron regulatory proteins. *Blood* **2011**, *118*, 168–179. [CrossRef] [PubMed]
30. Wilkinson, N.; Pantopoulos, K. The IRP/IRE system in vivo: Insights from mouse models. *Front. Pharmacol.* **2014**, *5*, 176. [CrossRef] [PubMed]
31. Anderson, S.A.; Nizzi, C.P.; Chang, Y.I.; Deck, K.M.; Schmidt, P.J.; Galy, B.; Damnernsawad, A.; Broman, A.T.; Kendziorski, C.; Hentze, M.W.; et al. The IRP1-HIF-2α axis coordinates iron and oxygen sensing with erythropoiesis and iron absorption. *Cell Metab.* **2013**, *17*, 282–290. [CrossRef] [PubMed]
32. Anderson, E.R.; Taylor, M.; Xue, X.; Ramakrishnan, S.K.; Martin, A.; Xie, L.; Bredell, B.X.; Gardenghi, S.; Rivella, S.; Shah, Y.M. Intestinal HIF2α promotes tissue-iron accumulation in disorders of iron overload with anemia. *Proc. Natl. Acad. Sci. USA* **2013**, *110*, 4922–4930. [CrossRef] [PubMed]
33. Anjali, N.; Julio, C.R.; Poorna, S.; Sudipa, G.-R.; Ruth, A.S.; Timothy, L.S.; Richard, K.B.; Caroline, C.P. Activation of the HIF prolyl hydroxylase by the iron chaperones PCBP1 and PCBP2. *Cell Metab.* **2011**, *14*, 647–657.
34. Shah, Y.M.; Matsubara, T.; Ito, S.; Yim, S.H.; Gonzalez, F.J. Intestinal hypoxia-inducible transcription factors are essential for iron absorption following iron deficiency. *Cell Metab.* **2009**, *9*, 152–164. [CrossRef] [PubMed]
35. Taylor, K.M.; Morgan, H.E.; Johnson, A.; Nicholson, R.I. Structure-function analysis of a novel member of the LIV-1 subfamily of zinc transporters, ZIP14. *FEBS Lett.* **2005**, *579*, 427–432. [CrossRef] [PubMed]
36. Zhao, N.; Gao, J.; Enns, C.A.; Knutson, M.D. ZRT/IRT-like protein 14 (ZIP14) promotes the cellular assimilation of iron from transferrin. *J. Biol. Chem.* **2010**, *285*, 32141–32150. [CrossRef] [PubMed]
37. Gao, J.; Zhao, N.; Knutson, M.D.; Enns, C.A. The hereditary hemo-chromatosis protein, HFE, inhibits iron uptake via down-regulation of Zip14 in HepG2 cells. *J. Biol. Chem.* **2008**, *283*, 21462–21468. [CrossRef] [PubMed]
38. Liuzzi, J.P.; Aydemir, F.; Nam, H.; Knutson, M.D.; Cousins, R.J. Zip14 (Slc39a14) mediates non-transferrin-bound iron uptake into cells. *Proc. Natl. Acad. Sci. USA* **2006**, *103*, 13612–13617. [CrossRef] [PubMed]
39. Pinilla-Tenas, J.J.; Sparkman, B.K.; Shawki, A.; Illing, A.C.; Mitchell, C.J.; Zhao, N.; Liuzzi, J.P.; Cousins, R.J.; Knutson, M.D.; Mackenzie, B. Zip14 is a complex broad-scope metal-ion transporter whose functional properties support roles in the cellular uptake of zinc and nontransferrin-bound iron. *Am. J. Physiol. Cell Physiol.* **2011**, *301*, 862–871. [CrossRef] [PubMed]
40. Supak, J.; Chia-Yu, W.; Richard, C.; Wei, Z.; Alan, C.; Thomas, B.; Jae-Sung, K.; Shintaro, H.; Toshiyuki, F.; Mitchell, D.K. SLC39A14 is required for the development of hepatocellular iron overload in murine models of hereditary hemochromatosis. *Cell Metab.* **2015**, *22*, 138–150.
41. Supak, J.; Adenike, A.; Elizabeth, P.; Ralf, W.; Shintaro, H.; Toshiyuki, F.; Genesys, G.; Jessica, S.; Armin, G.; Christopher, J.; et al. SLC39A14 deficiency alters manganese homeostasis and excretion resulting in brain manganese accumulation and motor deficits in mice. *Proc. Natl. Acad. Sci. USA* **2018**, *115*, 1769–1778.

42. Aydemir, T.B.; Kim, M.; Kim, J.; Colon-Perez, L.M.; Banan, G.; Mareci, T.H.; Febo, M.; Cousins, R.J. Metal transporter Zip14 (Slc39a14) deletion in mice increases manganese deposition and produces neurotoxic signatures and diminished motor activity. *J. Neurosci.* **2017**, *37*, 5996–6006. [CrossRef] [PubMed]
43. Xin, Y.; Gao, H.; Wang, J.; Qiang, Y.; Imam, M.U.; Li, Y.; Wang, J.; Zhang, R.; Zhang, H.; Yu, Y.; et al. Manganese transporter Slc39a14 deficiency revealed its key role in maintaining manganese homeostasis in mice. *Cell Discov.* **2017**, *3*, 17025. [CrossRef] [PubMed]
44. Illing, A.C.; Shawki, A.; Cunningham, C.L.; Mackenzie, B. Substrate profile and metal-ion selectivity of human divalent metal-ion transporter-1. *J. Biol. Chem.* **2012**, *287*, 30485–30496. [CrossRef] [PubMed]
45. Wang, C.Y.; Jenkitkasemwong, S.; Duarte, S.; Sparkman, B.K.; Shawki, A.; Mackenzie, B.; Knutson, M.D. ZIP8 is an iron and zinc transporter whose cell- surface expression is up-regulated by cellular iron loading. *J. Biol. Chem.* **2012**, *287*, 34032–34043. [CrossRef] [PubMed]

MDPI
St. Alban-Anlage 66
4052 Basel
Switzerland
www.mdpi.com

Nutrients Editorial Office
E-mail: nutrients@mdpi.com
www.mdpi.com/journal/nutrients

www.ingramcontent.com/pod-product-compliance
Lightning Source LLC
Chambersburg PA
CBHW062036220926
51850CB00010B/2566
* 9 7 8 3 0 3 6 5 9 6 4 0 2 *